CONTEMPORARY PHYSICS

PROCEEDINGS OF THE INTERNATIONAL SYMPOSIUM

CONTEMPORARY PHYSICS

PROCEEDINGS OF THE INTERNATIONAL SYMPOSIUM

NATIONAL CENTRE FOR PHYSICS

ISLAMABAD, PAKISTAN 26 – 30 MARCH 2007

edited by

Jamil Aslam

Institute of High Energy Physics, China & COMSATS Institute of Information Technology, Pakistan

Faheem Hussain

Lahore University of Management Sciences, Pakistan

Riazuddin

National Centre for Physics, Pakistan

World Scientific

NEW JERSEY • LONDON • SINGAPORE • BEIJING • SHANGHAI • HONG KONG • TAIPEI • CHENNAI

Published by

World Scientific Publishing Co. Pte. Ltd.
5 Toh Tuck Link, Singapore 596224
USA office: 27 Warren Street, Suite 401-402, Hackensack, NJ 07601
UK office: 57 Shelton Street, Covent Garden, London WC2H 9HE

British Library Cataloguing-in-Publication Data
A catalogue record for this book is available from the British Library.

CONTEMPORARY PHYSICS
Proceedings of the International Symposium

ISBN-13 978-981-281-893-5
ISBN-10 981-281-893-6

Printed in Singapore by World Scientific Printers

Preface

These are the proceedings of the International Symposium on Contemporary Physics organised by the National Centre for Physics (NCP) in Islamabad, Pakistan, from 26 to 30 March, 2007. The Symposium was organised to celebrate the completion of the new campus of the NCP, an important milestone in the history of the NCP.

Since the NCP is set up to promote all aspects of physics research and teaching in Pakistan it was decided to hold a meeting which would review the current status of research in major frontline areas of physics. The areas of research reviewed in this symposium were Condensed Matter Physics, Particle Physics, Quantum Optics, Quantum Computing and Laser Physics, Nano-sciences, Synchrotron Radiation, Relativity, Astrophysics and Cosmology, Plasma Physics. Leading physicists reviewed these frontier areas of physics. The total number of participants, including 23 speakers, was 271, mostly from Pakistan. It was heartening to see the huge participation of a large number of young students and researchers from all over Pakistan. In fact the number of requests for participation was very high and the organising committee had to restrict the number because of financial and logistical restrictions.

We were very fortunate to have the honour of hosting Dame Louise Johnson, the widow of Prof. Abdus Salam, as one of the distinguished speakers. Prof. Salam's dream was to have an active physics research centre in Pakistan and we are sure that he would have been quite pleased by the progress being made by the NCP.

We would like to thank, above all, the speakers who with their well prepared and well presented talks made the Symposium a memorable event. Never before in Pakistan have we had the occasion to gather such a distinguished faculty on such a broad spectrum of subjects of topical interest.

The symposium was a great success. It was the NCP's major step towards fulfilling one of its objectives, namely inviting eminent physicists from all over the world to share their knowledge and expertise with relatively less-knowledgeable and younger scientists from developing countries. To make this possible financial and institutional support was necessary. The NCP provided the primary institutional support. We are most grateful to Prof. Qasim Jan, Vice-Chancellor of Quaid-i-Azam University, for providing institutional support including use of the Geophysics and Physics Auditorium. The Symposium would not have been a success without the sponsorship and generous financial support of the Abdus Salam International Centre for Theoretical Physics (ICTP), the Higher Education Commission (HEC) and the Pakistan Academy of Sciences. We are grateful to both these organisations for their support. We would also like to thank the members of the International Advisory Committee, the International Steering Committee and the Local Steering Committee for valuable advice.

Without the help of the staff of the NCP the conference would not have been a success. Our special thanks go to Ubaidullah Khalid, Gulzaman Khan, Kashif Sarfraz Khan, Mansoor Sheikh, Ashfaq Ahmad and the staff of the computer section. Special thanks go also to all the students of the NCP (far too many to name here) who voluntarily did all the tough legwork to make the running of the conference a great success.

M. Jamil Aslam
Faheem Hussain
Riazuddin

INTERNATIONAL ADVISORY COMMITTEE

R. Aymer; CERN, Geneva
S. J. Brodsky; SLAC, Stanford
K. Cheetham; UC, Santa Barbra
J. Ellis; CERN, Geneva
Xiangdong Ji; MIT, Cambridge
T. D. Lee; Columbia, New York
M. A. H. Maccallum; QMUL, London
S. Mahajan; Texas A&M, College Station
L. Maiani; La Sapienza, Rome
R. Penrose; Oxford
N. R. Rao, Jawaharlal Nehru Centre, Bangalore
R. Ruffini; La Sapienza, Rome
H. Schopper; CERN, Geneva
M. Scully; Texas A&M, Princeton
K. R. Sreenivasan; ASICTP, Trieste
Townes; UC, Berkeley
S. Ting; CERN, Geneva
F. Wilczek; MIT, Cambridge
M. Wortis; Simon Fraser, Burnaby

INTERNATIONAL STEERING COMMITTEE

Ahmed Ali; DESY, Hamburg
Viqar Husain; New Brunswick, Fredericton
Zahid Hussain; UC, Berkeley
Asif Khan; South Carolina, Columbia
Talat Rahman; Central Florida, Orlando
Qaisar Shafi; Delaware, Newark
Ismat Shah; Delaware, Newark
Suhail Zubairy; Texas A&M, College Station

LOCAL STEERING COMMITTEE

Pervez Hoodbhoy; QAU, Islamabad
Hafeez Hoorani; NCP, Islamabad
Faheem Hussain; CIIT/NCP, Islamabad (Co-chair)
M. Zafar Iqbal; QAU, Islamabad
Mujahid Kamran; PU, Lahore
Asghar Qadir; CAMP/NUST, Rawalpindi
Riazuddin; NCP, Islamabad (Chair)
Abdullah Sadiq; GIKI, Topi
Hamid Saleem; PINSTECH, Islamabad
Hassan Shah; GCU, Lahore

LIST OF SPEAKERS

A. Ali
J. Bergou
P. Fromherz
J. Hormes*
V. Husain
L. Johnson
R. Maartens*
H. Newman*
T. Rahman
M. Richardson*
H. Schopper
W. Schleich
I. Shah
P. K. Shukla
I. Siddiqi*
A. Smirnov*
K. R. Sreenivasan
G. Tonelli*
K. Von Klitzing*
H. Winick
Z. Yoshida
A. F. Zakharov
S. Zubairy*

---

* Their lectures could not be included in the proceedings.

Contents

SECTION I

SYNCHROTRON RADIATION AND APPLICATIONS

Physics and biology: applications of synchrotron radiation in biology

Louise N. Johnson

Laboratory of Molecular Biophysics, University of Oxford, Oxford OX1 3QU, UK
and
Diamond light Source, Harwell Science and Innovation Campus, Didcot, Oxon, OX11 0QX, UK

First I should like to record how happy and honoured I am to be able to attend the Symposium on Contemporary Physics to celebrate the opening of the new International Centre for Physics at Quaid-I-Azam University, Islamabad. This imaginative project, now made concrete with superb buildings for both science and accommodation of scientists, should promote new activities in Pakistan and internationally. It is a venture, which would have given great pleasure to Professor Abdus Salam, Nobel Laureate in Physics 1979, who throughout his scientific career fought to enhance scientific research and its applications in the developing world.

Early history of application of X-rays in biology

X-rays, part of the electromagnetic spectra, have short wavelengths (of the order of 0.1 nm), which make them suitable for probing molecular and atomic structures. Following the discovery of X-rays by Roentgen in 1896 in Wurzburg, Germany, there was an immediate application in medicine. The differential absorption of X-rays by the elements allowed images of biological tissues in which bone structures were contrasted against the soft tissue. Just seven years later in 1913 Lawrence Bragg observed diffraction from crystals of simple molecules, such as sodium chloride and other alkyl halides, zinc blende and iron pyrites (all cubic lattices), and derived the law for X-ray diffraction that bears his name. He showed that atomic structure could be deduced from the X-ray diffraction patterns using Fourier transform theory.

Work on crystals of biological molecules did not start until the 1930s and at that time these molecules proved much too complex to allow their structure solution. However by the 1950s much was understood about diffraction from naturally occurring fibres that were ordered in one dimension so that the outstanding diffraction patterns of DNA recorded by Rosalind Franklin and

Maurice Wilkins could be interpreted in terms of an underlying helical structure. At the same time Jim Watson and Francis Crick, working independently in Cambridge, were able to derive a model for DNA based on physical chemical principals, knowledge of the crystal structure of one of the component nucleotides, and knowledge of the chemistry and the ratios of the bases (adenine (A)/ thymine (T) = guanine (G)/ cytosine (C) = 1). The resulting double helical structure had profound implications. It showed how the genetic material, seemingly capable of infinite variety, was based upon a regular helical structure. The specificity between the component base pairs (A:T and C:G) explained the fidelity of DNA replication.

The first protein structure, myoglobin, was solved in 1959 by John Kendrew and his team at Cambridge and by 1962 Max Perutz had a sub-atomic resolution structure of haemoglobin, the iron containing protein that is responsible for the transport of oxygen in blood. In 1965 David Phillips and his team at the Royal Institution in London solved the second protein structure, that of lysozyme, at atomic resolution. This was a landmark structure. Lysozyne was an enzyme, a biological catalyst capable of destroying bacterial cell walls. Knowledge of the structure led to an understanding of how the enzyme promoted this chemical transformation. Thus information on the three dimensional arrangement of the atoms in a molecule led to an understanding of biological function.

Slowly the new field of protein crystallography began to take off. But structure determination was painfully slow. The field required advances in protein preparation, now achieved for many proteins through advances in recombinant DNA technology and the ability to express a foreign protein in a culture of bacterial or eukaryotic cells. It required advances in protein crystallisation, an area where there is still a need for improvements; in computers and computer graphics; in algorithms and new innovations for solving the fundamental phase problem of crystallography; but most of all it required improvements in X-ray sources. Crystals of biological molecules are usually more than 50% solvent and diffract poorly. They also suffer from radiation damage, damage that can be partly alleviated by cryo-cooling the crystals to liquid nitrogen temperatures (~100 K). The conventional laboratory based X-ray generators provided a beam of such low intensity that long exposures were required and significant radiation damage occurred during these long exposures.

Synchrotron radiation

Synchrotron radiation, the radiation produced when an electron is accelerated on a circular path (as described elsewhere in this volume), provided the answer for greater intensity of radiation for biological specimens. The first synchrotron X-ray picture of a biological specimen was obtained by Rosenbaum, Holmes and Witz working at DESY, Hamburg in 1971 (Rosenbaum et al, 1971). They observed diffraction from a muscle specimen with an exposure time of 12 mins compared to 24 h that was needed in the laboratory source. Later in 1979 I had direct experience of the advantages for synchrotron radiation for protein crystallography when my graduate student Enrico Stura recorded diffraction patterns from a glycogen phosphorylase crystal at the synchrotron LURE, near Paris. For phosphorylase crystals at LURE, 6 mins exposures gave patterns of much greater quality than had been obtained in the laboratory with 13 h exposures. At that time structural biologists were using synchrotrons in parasite mode. Synchrotrons had been constructed primarily for elementary particle physics and diffraction experiments could only be performed in the calm periods when elementary particle collisions were not being investigated.

In 1981 the first synchrotron source dedicated to production of radiation for diffraction, scattering and spectroscopic applications was opened at Daresbury in the UK. Later, several other sources were built, many driven by the advantages of such radiation for biological studies. Around 1994, third generation synchrotron sources were introduced that use the latest technology for electron acceleration and insertion devices to produce even more intense radiation to be delivered to the experimental stations. Such sources include the European Synchrotron Source at Grenoble, France, the Advanced Photon Source at Argonne, USA, SPring 8 at Hyogo, Japan and the Swiss Light Source near Zurich, Switzerland. To date there are about 42 sources operational worldwide and a further 10 are under construction. About 20 of these have facilities for biological research.

In 2007 the UK commissioned its new third generation source, Diamond Light Source in Oxfordshire (www.diamond.ac.uk). This is a 3 GeV energy machine based on a 6-fold symmetry of 24 cells. The circumference is 561 m. Initially it will run with a current of 300 mA, rising to 500 mA in the future. Diamond has the lowest emittance of any source worldwide. The emittance describes the size and divergence of the electron beam. For Diamond the emittance is 2.7 nm rad (horizontal). This gives a very fine electron beam whose quality is reflected in the quality of X-rays produced.

Diamond Light Source has seven beam lines commissioned that include three beam lines for macromolecular crystallography, and beam lines for nanoscience, extreme conditions, materials and magnetism, and micro spectroscopy. In the years 2007-2011 these will be joined by a further 15 beam lines that reflect user scientific interests in both the life sciences and the physical sciences. In the life sciences these will include beam lines for non-crystalline diffraction, circular dichroism, infrared spectroscopy, and coherence and imaging.

In synchrotron science, very small samples are being observed, many of the orders of microns. This means that the synchrotron floor must be stable to within 1 micron. At Diamond this was achieved by sinking 1500 support columns through 15 m of chalk and clay to reach the hard rock. The experimental floor of the synchrotron is supported on these columns. There is a small team working at Diamond, who are expert in vibrational measurements. On the night of 8th October 2005, the tri-axial sensors were located in the office building and not the experimental floor. These sensors picked up the earthquake in Pakistan at 4.00 am. The tremors took 9 mins to travel through the earth to reach Oxfordshire in the UK 3,800 miles away. The signals indicate how we are all globally connected. One could only wish that the signals could be picked up in some way before the collisions that gave rise to the earthquake.

Applications of synchrotron radiation in macromolecular crystallography

The implications of synchrotron radiation for macromolecular crystallography have been enormous. By February 2007, there were 35,361 protein structures deposited in the Protein Data Bank (http://www.rcsb.org/pdb). Not all of these are unique structures. Several are the same protein from a different organism or the same protein with a ligand bound. It is estimated about 15,803 structures are unique (<95% sequence identity) or 8448 (<30% sequence identity). The escalation of numbers compared with the 2 structures that were available in 1965 is phenomenal. Over 80% of these structures are determined with synchrotron radiation.

Synchrotron radiation has three major advantages for macromolecular crystallography. Firstly, the radiation is orders of magnitude brighter than can be obtained with a laboratory source. Secondly, the radiation has a very small divergence allowing small crystals to be irradiated with an appropriate source size. Thirdly, the radiation is tuneable over the range of electromagnetic wavelengths from infrared to hard X-rays. This allows the radiation wavelength to be tuned to the natural absorption wavelength of elements in the crystal. Near these wavelengths anomalous scattering is observed and this scattering can be used to produce information on the phase of the diffracted beam. Frequently the sulphur containing methionine residues in protein molecules can be substituted by seleno-methionine. The selenium absorption edge is approximately 0.98 Å, a wavelength that can be conveniently reached with synchrotron radiation. Observation of anomalous scattering around this edge has led to the *ab initio* structure determination of many proteins. However this method requires that the protein can be expressed in *E. coli* bacterial cells that are modified so that they do not make their own methionine but instead incorporate the substituted seleno-methionine into the proteins.

Achievements

As indicated above, the application of synchrotron radiation in biology has been enormously productive (Blundell, 2005). The structures have given biological insights into: the functioning of enzymes that utilise a variety of chemical mechanisms; the recognition between protein molecules important for biological signalling; the recognition between protein and nucleic acids, DNA and RNA, important for understanding the regulation of gene expression and for DNA repair in response to radiation damage; into membrane proteins and how they function in the midst of the impermeable lipid bilayer; into the structure of viruses and their assembly from component protein molecules; into macromolecular assemblies such as the ribosome, the protein synthetic machine that decodes messenger RNA (mRNA) with the help of transfer RNA (tRNA) to accurately synthesise proteins encoded in the genome; into antibodies and T cell receptors and how they mediate the immune response; and many more examples.

In the last 20 years there have been four Nobel prizes for work in this field. In 1989 Johan Deisenhofer, Robert Huber and Hartmut Michel were recognised for their work on the bacterial reaction centre (Deisenhofer et al, 1985). The photosynthetic reaction centre from the purple bacteria *Rhodopseudomonas virides* is similar to photosystem II of plants. It is a membrane bound protein that creates an electrical gradient in response to absorption of light. The work showed how within this large protein complex the light absorbing chromophores were arranged to allow the transfer of electron from one side of the membrane to the other. In 1997 John Walker was honoured for his work on the F1-ATPase structure and its function (Abrahams et al, 1994). This large protein assembly is responsible for the synthesis of ATP, the energy currency of the cell. The structure confirmed a previously proposed binding-change model whereby, in response to a hydrogen ion gradient, the three subunits arranged around a central stalk protein respond with slightly different structures depending upon whether they are empty, bound to the precursor of ATP, or bound to ATP. In 2004 Rod McKinnon was honoured for his work on potassium channels (Doyle et al, 1998). The cell carefully regulates the concentration of ions. In response to neuronal signals potassium ions flow into the cell. The channel discriminates between sodium and potassium. The structural studies on the potassium channel showed the molecular basis for this discrimination and how ions could be selectively transported across the impermeable membrane lipid bilayer. The latest honour was the award of the Nobel Prize to Roger Kornberg in 2006 for his work on RNA polymerase (Cramer et al, 2001). This large protein complex carries out the essential step in the translation of DNA to mRNA and is regulated by several proteins to ensure fidelity of transcription.

Drug design

Structural biology has become an important component of drug discovery. In the search for new drugs, protein molecules are identified as possible targets either because they are known to be defective in the diseased state or because they control processes that are defective in disease such as the inflammatory response or unwanted cell growth in cancer. Knowledge of the structure of these target proteins can demonstrate definitively how a lead compound binds to the target protein and exerts its effects. This knowledge can then be used to modify the compound to provide a molecule of greater potency. Structure guides where and how a compound can be modified and where it should not be modified. Table I shows a number of compounds that are in the clinic and where structure has been informative in demonstrating the mechanism of action of the drug. Recent examples have included a drive for new anti HIV drugs (Brigo et al, 2005; Ghosh et al, 2006).

Protein kinases are frequent drug targets (Noble et al, 2004). These enzymes form part of a large family of proteins (518 members in the human genome). They function to phosphorylate target proteins and are key players in signalling pathways that control gene expression, cell growth and cell division, hormone response, inflammation and a whole host of other functions. They are principally targets for anti-cancer drugs. In work in my own laboratory we used the structure of the protein kinase cyclin dependent protein kinase 2 (CDK2) to design a compound that had high affinity (Davies et al, 2002). Our starting point was a lead compound O^6-cyclohexylmethoxyguanine that had micro molar affinity for the enzyme and inhibited the kinase action. Through knowledge of the structure we were able to design a compound, O^6-cyclohexylmethoxy-2-(4'sulfamoylanillino)guanine, that had nanomolar affinity, three orders of magnitude tighter. However it is a long road from a nanomolar affinity compound to a drug that is the clinic. The compounds must satisfy toxicology requirements, they need to be able to get into cells, they must be absorbed and excreted at suitable rates and time scales, and they must meet a host of requirements for stability.

To date there are a number of kinase inhibitors that have successfully met these requirements and are in the clinic as effective drugs (Table II). Each of these is specific for just a few protein kinases out of the 518 protein kinases that are known to be present in the human genome. Structure has informed on the basis for this specificity and mechanism of action. For example Glivec (or Gleevec), one of the most successful drugs for the treatment of chronic myeloid leukaemia, inhibits the kinase Abelson tyrosine kinase. It does this by stabilising an inactive conformation of the kinase and is selective because such conformations cannot be achieved by other kinases (Nagar et al, 2002). With long-term administration of the drug, patients with resistance have emerged. Sequencing of the Abl tyrosine kinase from these patients indicated with the help of structure how resistance had occurred and led to new compounds to overcome resistance (Shah et al, 2002; Young et al, 2006).

A further fruitful attack on cancer has come from raising monoclonal antibodies that recognise and inactivate the hormonal receptors signalling growth that are often defective in cancer. Some of these are summarised in Table II. Again structure of these therapeutic antibodies with their target proteins has demonstrated mechanism of action, mechanisms that are different for each of the three antibodies listed and which would have been difficult to predict without the benefit of structure.

Future

With this wealth of information, what remains to be done in macromolecular crystallography?

- Crystallographers are now tackling **larger and more complex assemblies and machines** hoping to understand how the assembly as a whole can confer properties not encompassed in the individual components, as for example recently demonstrated in the structure of the intact 70S ribosome (Selmer et al, 2006).

- Because of the power of synchrotron sources and microfocus optics on some beam lines, crystallographers can now work with **very small crystals** (even < 10 microns). Many interesting proteins such as those from membrane proteins give very small crystals that diffract poorly. Indeed **membrane proteins** because they are difficult to purify and crystallise in their functionally active state are very much underrepresented in the Protein Data Bank. Of the 35,000 structures deposited only about 130 are membrane proteins and yet from knowledge of the human genome about 30% of the proteins encoded are membrane proteins. However progress is being made with membrane proteins (Byrne & Iwata, 2002; Horsefield et al, 2003; Iwata & Barber, 2004).
- **How many ways can a protein fold**? It appears that we are now approaching almost completion in the recognition of protein folds. About 12% of new protein structures (that exhibit less than 30% sequence identity to an existing structure) have a new fold. As more structures accumulate we might expect to have a complete dictionary of protein folds. This will assist in tackling one of the most intractable problems in theoretical structural biology. How to predict a protein fold from an amino acid sequence? A few years ago there was a significant advance in which Baker and his colleagues (Kuhlman et al, 2003) in solving an inverse folding problem. They showed that a novel fold could be designed and a sequence constructed that would adopt this fold.
- There will continue to be significant contributions from synchrotron radiation in macromolecular crystallography in **structure based drug design**. Many pharmaceutical companies are adopting the fragment-based approach in the search for new compounds and this is dependent on structures being available to indicate where the fragments bind and how they might be linked to produce a lead compound (Blundell & Patel, 2004; Congreve et al, 2005; Howard et al, 2006).
- The high brightness synchrotron sources allow **time resolved experiments** where intermediates of short life times can be detected. Such an application was an experiment in which the dissociation of carbon monoxide from the heme in myoglobin was tracked (Chu et al, 2000). Such experiments are not easy. They require that an interesting event can take place within the protein in the crystalline lattice on time scales of the order of milliseconds to microseconds.
- There is a need to build from our understanding of biology at the molecular level, through the structures of individual components, to their functioning

in the whole cell. In the first instance this has involved links between imaging by **X-ray crystallography and electron microscopy**.

- Finally there are proposals using the new sources of **free electron lasers** to be able to image molecules, assemblies or cells without the need for crystallisation. Free electron laser X-ray sources will provide a beam that is orders of magnitude greater (perhaps 10^{10} times) than can be achieved with storage ring synchrotron source. Such sources will not be available until 2009 but very recently promising proof of principal experiments have been reported (Chapman et al, 2007).

Other applications of synchrotron radiation in biology

Non-crystalline diffraction

Many biological samples are naturally semi-ordered in one dimension such as DNA, muscle, skin, cornea, amyloid fibres etc. Diffraction from these specimens can give information not at the atomic level but in terms of organisation that can be informative. For example the study of collagen in the cornea can be informative for defects that arise in diseases like glaucoma (Aghamohammadzadeh et al, 2004; Boote et al, 2005). Or the study of the collagen from skin in ancient parchments can give information on the state of preservation and to some extent the origin of the sample. In addition solution X-ray scattering studies can give information on size and shape of macromolecules and how they respond in different assemblies. Such information is becoming important as macromolecular crystallographers tackle large assemblies.

Circular dichroism

Circular dichroism (CD) in the visible and ultraviolet spectral regions has long been a laboratory based spectroscopic technique for assessing the folded state of proteins and other biological molecules and estimation of their secondary structure. Latterly the method has applications in determining ligand binding affinities and more detailed characterisation of conformational changes in response to ligand or environmental changes. The major advantages of the synchrotron light source for CD are: the higher photon flux throughout the whole spectra and hence improved signal to noise allowing work with more dilute samples; the possibility of a wider wavelength range at short wavelengths in the extended vacuum-UV region to 165 nm for aqueous samples and to 130 nm for solid samples; and the smaller image size at the focal plane (about 1 mm x 1 mm) allowing the use of small sample volumes (Miles & Wallace, 2006).

Infrared Microspectroscopy

Observation of the infrared (IR) vibrational absorption peaks characteristic of specific bond arrangements can be informative for a number of studies. IR imaging of cells can exploit the particular absorption vibrational frequencies of different chemical groups to provide differential contrast for example by protein, lipid or DNA. IR radiation is non-ionising and non-destructive. The technique has been extended to monitoring the response of tumour cells to chemotherapy, with potential benefit for diagnosis and treatment and a number of other studies of medical relevance (Dumas et al, 2007; Yang et al, 2005). Applications in environmental sciences and geobiology have involved measurements on biomineralisation and detection of environmental pollutants. In archeometry there have been applications in analysis of paints, fabric sections and ancient surface structures using very small samples. In food science FT-IR can be used to monitor changes following heat induced denaturation and properties of water molecules. The increased intensity of the synchrotron radiation compared to laboratory based sources and the ease of wavelength tuning will contribute to the advantages of use of synchrotron radiation in these studies.

Soft X-ray imaging

Imaging in biology using synchrotron radiation can be divided into two different approaches. Firstly it is possible to construct an X-ray microscope using Fresnel zone plate lenses to focus X-rays on the specimen as a condenser lens and then to collect the scattered X-rays to form a direct image. The resolution is limited by the quality of the Fresnel zone plates and by the diameter of the outer ring. Using 2.4 nm radiation this method has been applied with great effect to image yeast cells (Larabell & Le Gros, 2004). Moreover tomography is relatively easy to apply to generate three-dimensional images. The resolution, between 20-60 nm, is not as good as can be achieved with the electron microscope where resolution for cells of 5-6 nm are possible with cryo-electron tomography (Medalia et al, 2002) but in the X-ray microscope thicker specimens can be used (perhaps as great as 10 microns). Both the X-ray microscope and the electron microscope imaging techniques suffer from the disadvantage that the specimen must be frozen in order to avoid radiation damage. This compares unfavourably with current confocal optical microscopy where live specimens can be imaged. Nevertheless X-ray imaging is an exciting development that is likely to expand in the next few years and to deliver results for biology in the intermediate resolution range.

A second approach in imaging has been to use a lensless system that exploits the coherence of synchrotron radiation. Here the X-ray scattering from the sample is recorded, essentially the molecular transform of the object. Reconstruction to an image can be achieved using the over-sampling theorem to deduce the phases of the scattered X-rays. An example of this approach has been the imaging of *E. coli* cells at 30 nm resolution using radiation of wavelength 2.0 Å (Miao et al, 2003). The authors were able to show good correspondence between the X-ray images and those obtained by light microscopy.

Medical beam lines

There have been a number of different approaches to exploit synchrotron sources for medicine. Several synchrotron sources have dedicated medical beam lines where for example direct imaging with differential contrast is used to image tissues such as in cancer (Bravin et al, 2007) or the lungs of an animal as it takes its first breath using very low doses that do not lead to damage (Hooper et al, 2007). Applications have included coronary angiography in which arteries of human volunteers were imaged using a contrast agent (iodine or gadolinium) injected intravenously (Bertrand et al, 2005). With the synchrotron source, very much lower concentrations of contrast agents were possible. The results showed much clearer imaging than achieved with conventional angiography. However clinicians do not wish to send sick patients to a synchrotron source for these sorts of analyses but prefer to have the facilities available in the local hospital. Small bench-top based synchrotrons with the benefit of tuneable radiation are becoming available that might allow development in this area in the future.

An area that appears to be of considerable promise is in radiation therapy for treatment of cancers where the unique features of synchrotron radiation could give benefits not available elsewhere. Microbeam radiation therapy using small beams 10-30 microns wide spaced by 200 microns in a parallel array have been used to treat brain tumours in rats and pigs. Of greater significance is the combination of chemotherapy and radiation therapy. Results have shown that tumours first dosed with the anticancer drug platinum diaminodichloride (cisplatin) and then irradiated with a microbeam at wavelength corresponding to the K absorption edge of platinum (78 keV) gave significant reduction in glioma tumour size (Biston et al, 2004). This approach, known as contrast enhanced synchrotron stereotactic radiotherapy, is now going into Phase I clinical trials with human patients at the ESRF, Grenoble.

Concluding remarks

This discussion is not inclusive of all applications of synchrotron radiation in biology. For example I have not discussed the use of microspectroscopy or EXAFS (extended fine edge absorption spectroscopy) for the analysis of chemicals and metals in specimens. However the intention has been to show how the synchrotron light in Bragg's description is 'a messenger carriying a story about the form of the object'. There is much that can be delivered with existing technology and several areas that remain to be developed and which show exciting promise for the future.

Table I. Marketed drugs for which structural biology has contributed information

Drug	Compound	Company	Disease target	Protein target	PDB entry
Gleevec	Imatinib	Novartis	Chronic myeloid leukemia Gastrointestinal stroma tumours	Abl Tyrosine kinase C-Kit PDGFR	1XBB
Herceptin	Trastuzumab	Genentech	Breast cancer	Her2 receptor	1N8Z
Lipitor	Atorvastatin	Pfizer	High cholesterol	HMG (3-hydroxyl-3-methylglutaryl) CoA reductase	1HWK
Avandia	Rosiglitazone	GSK	Type 2 diabetes	Peroxisome proliferator-activated receptor (PPARγ)	2PRG
Actonel	Risedronate	Proctor & Gamble	Osteoporosis	Farnesyl diphosphate synthase	1YV5
Casodex	Bicalutamide	AstraZeneca	Prostate cancer	Androgen receptor	1E3G
Norvir	Ritonavir	Abbott	HIV	HIV protease	1HXW
Relenza	Zanamivir	GSK	Influenza	Influenza neuraminidase	1A4G

Table II. Protein kinase targeted drugs

a) Protein kinase inhibitors

Drug (date approved)	Target protein kinase	Disease
Fasudil (1999)	Rho-kinase	Cerebral vasospasm
Gleevec STI-571 (2001)	Abl tyrosine kinase c-Kit, PDGFR	Chronic myeloid leukemia Gastrointestinal stromal tumours
Iressa ZD-1839 (2004)	EGFR tyrosine kinase	Non-small cell lung cancers (esp. adenocarcinomas)
Tarceva OSI 774 (2004)	EGFR tyrosine kinase	Non-small cell lung and pancreatic carcinomas
Sorafenib BA 43-9006 (2006)	B-RAF, VEGFR, PDGFR, FLT3	Renal cell carcinoma
Sunitinib SU11248 (2006)	VEGFR, PDGFR, FLT3, c-Kit	Renal cell carcinoma, GIST

b) Protein kinase and growth factor therapeutic antibodies

Therapeutic antibody (date approved)	Target protein	Disease
Herceptin (Trastuzumab) (2002)	Her2 EGF receptor	Breast Cancer
Erbitux (Cetuximab) (2004)	EGF receptor	Metastatic colorectal cancer
Avastin (Bevacizumab) (2004)	VEGF	Metastatic colorectal cancer

Abrahams JP, Leslie AG, Lutter R, Walker JE (1994) Structure at 2.8 A resolution of F1-ATPase from bovine heart mitochondria. . *Nature* **370:** 621-628

Aghamohammadzadeh H, Newton RH, Meek KM (2004) X-ray scattering used to map the preferred collagen orientation in the human cornea and limbus. *Structure* **12:** 249-256

Bertrand B, Esteve F, Elleaume H, Nemoz C, Fiedler S, Bravin A, Berruyer G, Brochard T, Renier M, Machecourt J, Thomlinson W, Le Bas JF (2005) Comparison of synchrotron radiation angiography with conventional angiography for the diagnosis of in-stent restenosis after percutaneous transluminal coronary angioplasty. *European heart journal* **26:** 1284-1291

Biston MC, Joubert A, Adam JF, Elleaume H, Bohic S, Charvet AM, Esteve F, Foray N, Balosso J (2004) Cure of Fisher rats bearing radioresistant F98 glioma treated with cis-platinum and irradiated with monochromatic synchrotron X-rays. *Cancer research* **64:** 2317-2323

Blundell TL (2005) Advances in the uses of synchrotron X-ray and proton beams in structural biology. *Progress in biophysics and molecular biology* **89:** 121-123

Blundell TL, Patel S (2004) High-throughput X-ray crystallography for drug discovery. *Current opinion in pharmacology* **4:** 490-496

Boote C, Dennis S, Huang Y, Quantock AJ, Meek KM (2005) Lamellar orientation in human cornea in relation to mechanical properties. *Journal of structural biology* **149:** 1-6

Bravin A, Keyrilainen J, Fernandez M, Fiedler S, Nemoz C, Karjalainen-Lindsberg ML, Tenhunen M, Virkkunen P, Leidenius M, von Smitten K, Sipila P, Suortti P (2007) High-resolution CT by diffraction-enhanced x-ray imaging: mapping of breast tissue samples and comparison with their histo-pathology. *Physics in medicine and biology* **52:** 2197-2211

Brigo A, Mustata GI, Briggs JM, Moro S (2005) Discovery of HIV-1 integrase inhibitors through a novel combination of ligand and structure-based drug design. *Medicinal chemistry (Shariqah, United Arab Emirates)* **1:** 263-275

Byrne B, Iwata S (2002) Membrane protein complexes. *Current opinion in structural biology* **12:** 239-243

Chapman HN, Hau-Riege SP, Bogan MJ, Bajt S, Barty A, Boutet S, Marchesini S, Frank M, Woods BW, Benner WH, London RA, Rohner U, Szoke A, Spiller E, Moller T, Bostedt C, Shapiro DA, Kuhlmann M, Treusch R, Plonjes E, Burmeister F, Bergh M, Caleman C, Huldt G, Seibert MM, Hajdu J (2007) Femtosecond time-delay X-ray holography. *Nature* **448:** 676-679

Chu K, Vojtchovsky J, McMahon BH, Sweet RM, Berendzen J, Schlichting I (2000) Structure of a ligand-binding intermediate in wild-type carbonmonoxy myoglobin. *Nature* **403:** 921-923

Congreve M, Murray CW, Blundell TL (2005) Structural biology and drug discovery. *Drug discovery today* **10:** 895-907

Cramer P, Bushnell D, Kornberg R (2001) Structural basis of transcription: RNA polymerase II at 2.8 angstrom resolution. *Sceince* **292:** 1863-1876

Davies TG, Bentley J, Arris CE, Boyle FT, Curtin NJ, Endicott JA, Gibson AE, Golding BT, Griffin RJ, Hardcastle IR, Jewsbury P, Johnson LN, Mesguiche V, Newell DR, Noble ME, Tucker JA, Wang L, Whitfield HJ (2002) Structure-based design of a potent purine-based cyclin-dependent kinase inhibitor. *Nature structural biology* **9:** 745-749

Deisenhofer J, Epp O, Miki K, Huber R, Michel H (1985) Structure of the protein subunits in the photosynthetic reaction centre of Rhodopseudomonas viridis at 3 A resolution. *Nature* **318:** 618-624

Doyle D, Morais Cabral J, Pfuetzner R, Kuo A, Gulbis J, Cohen S, Chait B, MacKinnon R (1998) The structure of the potassium channel: molecular basis of K+ conduction and selectivity. *Science (New York, NY)* **280:** 69-77

Dumas P, Sockalingum GD, Sule-Suso J (2007) Adding synchrotron radiation to infrared microspectroscopy: what's new in biomedical applications? *Trends in biotechnology* **25:** 40-44

Ghosh AK, Sridhar PR, Leshchenko S, Hussain AK, Li J, Kovalevsky AY, Walters DE, Wedekind JE, Grum-Tokars V, Das D, Koh Y, Maeda K, Gatanaga H, Weber IT, Mitsuya H (2006) Structure-based design of novel HIV-1 protease inhibitors to combat drug resistance. *Journal of medicinal chemistry* **49:** 5252-5261

Hooper SB, Kitchen MJ, Wallace MJ, Yagi N, Uesugi K, Morgan MJ, Hall C, Siu KK, Williams IM, Siew M, Irvine SC, Pavlov K, Lewis RA (2007) Imaging lung aeration and lung liquid clearance at birth. *Faseb J*

Horsefield R, Yankovskaya V, Tornroth S, Luna-Chavez C, Stambouli E, Barber J, Byrne B, Cecchini G, Iwata S (2003) Using rational screening and electron microscopy to optimize the crystallization of succinate:ubiquinone oxidoreductase from Escherichia coli. *Acta crystallographica* **59:** 600-602

Howard N, Abell C, Blakemore W, Chessari G, Congreve M, Howard S, Jhoti H, Murray CW, Seavers LC, van Montfort RL (2006) Application of fragment screening and fragment linking to the discovery of novel thrombin inhibitors. *Journal of medicinal chemistry* **49:** 1346-1355

Iwata S, Barber J (2004) Structure of photosystem II and molecular architecture of the oxygen-evolving centre. *Current opinion in structural biology* **14:** 447-453

Kuhlman B, Dantas G, Ireton GC, Varani G, Stoddard BL, Baker D (2003) Design of a novel globular protein fold with atomic-level accuracy. *Science (New York, NY)* **302:** 1364-1368

Larabell CA, Le Gros MA (2004) X-ray tomography generates 3-D reconstructions of the yeast, saccharomyces cerevisiae, at 60-nm resolution. *Molecular biology of the cell* **15:** 957-962

Medalia O, Weber I, Frangakis AS, Nicastro D, Gerisch G, Baumeister W (2002) Macromolecular architecture in eukaryotic cells visualized by cryoelectron tomography. *Science (New York, NY)* **298:** 1209-1213

Miao J, Hodgson KO, Ishikawa T, Larabell CA, LeGros MA, Nishino Y (2003) Imaging whole Escherichia coli bacteria by using single-particle x-ray diffraction. *Proceedings of the National Academy of Sciences of the United States of America* **100:** 110-112

Miles AJ, Wallace BA (2006) Synchrotron radiation circular dichroism spectroscopy of proteins and applications in structural and functional genomics. *Chemical Society reviews* **35:** 39-51

Nagar B, Bornmann WG, Pellicena P, Schindler T, Veach DR, Miller WT, Clarkson B, Kuriyan J (2002) Crystal structures of the kinase domain of c-Abl in complex with the small molecule inhibitors PD173955 and imatinib (STI-571). *Cancer research* **62:** 4236-4243

Noble ME, Endicott JA, Johnson LN (2004) Protein kinase inhibitors: insights into drug design from structure. *Science (New York, NY)* **303:** 1800-1805

Rosenbaum G, Holmes KC, Witz J (1971) Synchrotron radiation as a source for X-ray diffraction. *Nature* **230:** 434-437

Selmer M, Dunham CM, Murphy FVt, Weixlbaumer A, Petry S, Kelley AC, Weir JR, Ramakrishnan V (2006) Structure of the 70S ribosome complexed with mRNA and tRNA. *Science (New York, NY)* **313:** 1935-1942

Shah NP, Nicoll JM, Nagar B, Gorre ME, Paquette RL, Kuriyan J, Sawyers CL (2002) Multiple BCR-ABL kinase domain mutations confer polyclonal resistance to the tyrosine kinase inhibitor imatinib (STI571) in chronic phase and blast crisis chronic myeloid leukemia. *Cancer cell* **2:** 117-125

Yang Y, Sule-Suso J, Sockalingum GD, Kegelaer G, Manfait M, El Haj AJ (2005) Study of tumor cell invasion by Fourier transform infrared microspectroscopy. *Biopolymers* **78:** 311-317

Young MA, Shah NP, Chao LH, Seeliger M, Milanov ZV, Biggs WH, 3rd, Treiber DK, Patel HK, Zarrinkar PP, Lockhart DJ, Sawyers CL, Kuriyan J (2006) Structure of the kinase domain of an imatinib-resistant Abl mutant in complex with the Aurora kinase inhibitor VX-680. *Cancer research* **66:** 1007-1014

Sesame – a project to foster science and peace and its relavence for the region

Herwig Schopper
Chairman SESAME Council[*]

It is a great pleasure to thank Professor Riazuddin for inviting me to this symposium which is held on the occasion of the inauguration of the new National Centre for Physics. We all should congratulate him for this great achievement which will be so beneficial both for international cooperation and for the strengthening of science in Pakistan.

1. Why a synchrotron radiation facility?

SESAME (Synchrotronlight for Experimental Science and Application in the Middle East) is an international research centre for the Middle East and the Mediterranean Basin created under the umbrella of UNESCO following the model of CERN. It is the first international research organization in Muslim countries. It has two main objectives: Promote science and technology in the region and help to improve mutual understanding and building trust among people from different traditions, religions, political systems and races according to the UNESCO slogan "Science for Peace".

How can this task be achieved?

- Create a centre of excellence for interdisciplinary research with a competitive research facility
- Promote international cooperation
- Use scientific excellence as one of the main guiding principles
- Training of scientists, students and technicians (and even administrators)
- Attract scientists from the region working abroad and thus reverse at least partially the brain drain
- Promote the development of applications and high-tech industry.

[*] Invited talk at the International Symposium on Contemporary Physics, 26-30 March 2007, Islamabad.

The main facility will be an accelerator to produce synchrotron radiation. It will be an extremely strong light source with a spectral range from the infrared region to X-rays.

Apart from high intensities the light has special properties (very short pulses, polarized) offering research opportunities in many fields.
Individual beams for specific interests of potential users can be set up and the main research domains will be:

- Physics (mainly condensed matter)
- Material science
- Nanotechnology
- Molecular biology
- Archaeology
- Environmental studies
- Medical research.

Synchrotron Radiation sources (SR) have become ubiquitous tools for research and about 60 sources are in operation worldwide with more than 20.000 users. They exist not only in industrialised countries some of which operate more than one source, but also in developing countries like Brazil, China, India, Korea, Taiwan and Thailand. SR sources are cheaper and less complicated than neutron source (research reactors, spallation sources) and more versatile than lasers. There exist a few 'supermachines' in the world (investments of hundreds of million US dollar like ESRF at Grenoble, Spring 8 in Japan or the Argon Light Source in the USA), but smaller facilities like SESAME are competitive on many areas since they can do about 90 % of all the research requiring synchrotron radiation and only a small fraction needs the really big machines.

No SR source exists so far in the Middle East. Abdus Salam proposed such a facility at a Symposium held at Bahrain University in 1983 on the Future Outlook of the Arabian Gulf. I suggested a SR Lab for Saudi Arabia about 10 years ago when I gave talks in Ryadh. But nothing happened.

But now SESAME will be the first international facility in the region. However, I am convinced that it will not remain the only one. Large countries like Pakistan, Egypt, Turkey or Iran, will recognise the usefulness of such facilities and will eventually build their own machines. To realise such a national project, of course, the necessary funds of the order of about $ 50 million (assuming that the ground and other infrastructure would be available) must be provided. However, the more difficult condition for a successful project would be the creation of a crew competent in accelerator technique, beam design and other domains. The participation in the SESAME project offers an excellent opportunity to train scientists and engineers to gain such experience and this should be considered as a useful and necessary first step for a later national project. Even a small fraction (1 or 2%) of the funds which are spent for short-term problems (e.g. infrastructures for roads and water) would be sufficient for such long-term projects.

A SR source should be considered as an important element for the long-term development of a country or a region for the improving of the general conditions of the standard of life, foster education and reduce unemployment. In industrialised countries the employment in agriculture has gone down form some 60% to a few % whereas the employment in high-tech fields is increasing beyond 50%. This change in the sociological structure took more than 150 years in developed countries. Developing countries of today cannot afford to pass through such a long cycle and such a transition has to achieved in a much shorter time, perhaps 30 years.

2. The foundation of SESAME

SESAME has been created under the auspices of UNESCO, following the example of CERN, as an autonomous international organisation by a decision of the Executive Board of UNESCO in Spring 2002 that qualified SESAME as a "quintessential UNESCO project combining capacity building with vital peace-building through science" and described it as a "... model project for other regions....". UNESCO is the depository of the Statutes of SESAME. These Statutes came into force in April 2004 when six countries notified the Director-General, K. Matsuura, of their wish to join the Centre and their acceptance of the Statutes.

From 1999 until the Statutes came into force in April 2004, a number of countries participated in the work of an Interim Council (IC) that had been set up under F.Mayor, Director General of UNESCO at that time, to take the necessary measures to prepare the establishment of the Centre. They did so as either Members or Observers of the IC. Following the formal recognition of the Centre by UNESCO, most of the countries having participated in the IC confirmed their membership status and a permanent international Council became the governing body of SESAME. It is chaired by President H.Schopper, Germany and two Vice- Presidents, Dincer Ulkü, Turkey and Fawzy Elrefaie, Egypt. Each Member has one vote (two delegates) independent of its size or contribution.

The present eight Members of SESAME are Bahrain, Cyprus, Egypt, Israel, Jordan, Pakistan, the Palestinian Authority and Turkey (since Palestine is not yet a state one must speak of 'Members' and not 'Member Countries'). Some countries are still in the process of legalising their Membership e.g. Morocco, Oman, Iran and UAE. Iraq has recently asked to become a Member. France, Germany, Greece, Italy, Kuwait, the Russian Federation, Sweden, UAE, the U.K. and the U.S.A. are Observers. Spain and Japan are shortly expected to become Observers. Other countries have expressed interest (e.g. Libya, Yemen) in becoming Members. All countries in the region accepting the SESAME Statutes and willing to participate in SR research are welcome.

The basis for choosing Jordan as Host State was the assurance that all scientists of the world would have free access to the Centre, and the commitment by the Government of Jordan to provide the land, the funds for the construction of the building to house the facility, as well as some infrastructure (some existing buildings). H.M. Abdullah II, King of Jordan, strongly supports the project. Jordan and SESAME have signed a Seat Agreement guaranteeing diplomatic privileges to the organisation and its staff (including tax-free salaries and exemption from other taxes) similar to those which CERN has been given by its Host States. The site is located in a beautiful environment at an altitude of about 1000 m with an agreeable climate. It can be reached by car from Amman in about half an hour and Amman has an international airport which is important for future users of the facility.

Some countries have to acquire a new mentality having become a Member of an international laboratory. They have to learn that they are "owners" of the centre even if it is not located in their own country. Each Member, independent of its geographical location, has the same rights as far as the decision taking (definition of programmes, appointment of directors, etc), utilisation, recruitment of staff and industrial contracts are concerned. The host-state has rather more obligations than privileges. At CERN such a mentality has become tradition.

3. The SESAME facility

A conceptual design for the SESAME machine with a final energy of 2.5 GeV has finally been approved by Council after passing through a number of steps. It is based on the use of components of the BESSY I storage ring, a facility in West Berlin which stopped operation at the end of November 1999 for political and financial reasons (it had been decided to construct BESSY II in East Berlin). Following a suggestion by G.-A. Voss and H. Winick the German government donated the components of this machine provided SESAME covered the cost of $ 600.000 for its dismantling. The components were shipped from Germany to Jordan in 2002 and they are waiting to be installed. In particular the injection system consisting of a microtron and a booster synchrotron will be very valuable for SESAME and represent a saving of $ 6 to 7 million.

Originally it was foreseen to upgrade SESAME from the BESSY I energy of 0.8 GeV to 1 GeV with the main magnetic ring having a circumference of 100 m. However, the demands of the potential users made it soon very clear that a higher energy was required to provide an interesting and competitive facility. The main pressure came from biologists interested in structural molecular investigations. A new design was worked out for an energy of 2 GeV and a circumference of 116 m. Further discussions with the users lead to a final design with an energy of 2.5 GeV and a ring circumference of 133 m allowing beamlines with a length of up to 30 m. The layout of the accelerator hall is shown in Fig.1. Some of the 16 bending magnets will serve as sources for SR in particular for low photon energies. The main ring has 8 long straight sections for insertion devices like wigglers and undulators which are used to produce radiation with particular high intensities and special spectral features.

The construction work of the building that is to host the Centre started in August 2003 and the raw building was finished by the end of 2006 (Fig.2). The construction was finished in an astonishing short period in spite of considerable difficulties with the foundations and a crane spanning the complete hall. The installation of the injector system will start in 2007. The documents for the tendering procedures for the components of the main ring are prepared and the call for offers will soon start. The commissioning of the facility is planned for the end of 2010. SESAME will be a fully competitive 3 generation SR facility.

If this time schedule can be achieved the total period for the realisation of the project from the first ideas to the turn on of beamlines will be about 10 years. This is a very favourable time schedule in view of the difficult political and financial conditions. Even most of similar national projects took longer.

The Centre is managed by a Director (Minister Kh. Toukan, Jordan), an Administrative Director (H.Helal, Egypt, who has resigned after recently becoming minister), a Scientific Director (H.Hoorani following A. Baig, both from Pakistan) and a Technical Director (G. Vignola, Italy). It also has a core staff working on the up-grading of the machine. They are mainly accelerator specialists from the region who have been trained in synchrotron radiation laboratories in Europe and the United States.

Because of the way SESAME is being financed, it is difficult to quote a single figure for the total cost of the project. The ground, the building and other parts of the infrastructure have been provided by Jordan. The injector chain was a gift from the Germany. What remains is the upgrading of the main accelerator ring. Based on quotes from European industry an amount of EUR 15 million is neccesary. Since Members of SESAME are not accustomed to such expenses for scientific projects a request for these funds was submitted to the European Commission. After an evaluation by the EC EUR 1 million was granted in 2006 through a bilateral initiative programme for Jordan. The negotiations continue with the aim to receive more funds through a regional MEDA programme.. In order not to delay the project H.M. King Abdullah II has granted US $ 3.1 million in 2007 which are immediately available to place orders for the components of the main ring. To assure that SESAME remains a truly international project it is important to obtain additional financial contributions from other Members, the EC or other sponsors. The total funds involved are small compared to the funds available in the region for other infrastructure projects and therefore the main task is to convince decision makers that funds for the scientific infrastructure are an investment for the long-term welfare of society.

The operating cost is covered by contributions of the SESAME Members. The total budget amounts presently to about $ 1 million which is mainly used to pay salaries. With the occupation of the new building this budget has slowly to increase and is supposed to reach about $ 5 million in 2011 when the facility will be in operation.

To benefit as much as possible from international experience and competence the Council has established several international advisory committees: Beamlines Committee(chair Zahid Hussain, USA/Pakistan) to define the first generation of beamlines, to help to set up collaborations and later evaluating proposals; a Scientific Committee (chair Zehra Sayers, Turkey) to give advice on the long term science policy, e.g. upgrading of the machine; a Technical Committee (chair Albin F. Wrulich, Germany) to advice on the machine design, technical solutions and cost estimate; a Training Committee (chair Javad Rahighi, Iran) to help in organizing the training programme and the selection of candidates; a Finance Committee (Chair Minister H. Helal, Egypt) prepares the budget and the distribution of contributions among Members.

Beamlines for the first phase of exploitation have been identified in close cooperation with potential users (Fig.3). For this purpose the advisory committees have organized a number of users' meeting and workshops.

Beamlines will have to be financed from outside the normal budget. Funds for beamline equipment are being sought from other organisations like IAEA. Some Members of SESAME have expressed their intention to take the responsibility for particular beamlines and I hope that Pakistan will take the lead for one beamline. Some components will also be obtained from other laboratories, as SLAC (Stanford Linear Accelerator Centre) in the U.S. and LURE (Laboratoire pour l'Utilisation du Rayonnement Electromagnétique) in France which are donating components of beamlines from decommissioned machines.

4. Training

Training is one of the major objectives of SESAME, in particular

- the formation of young scientists and engineers of the region for the construction and operation of the machine,
- the training of scientists for the design, construction and operation of the beamlines,
- the education of potential users.

For this purpose fellowships and other support were made available by IAEA, ICTP, USA, Japan and some laboratories. From about 100 applicants from the region 20 technically minded persons were selected to spend a year or longer at European laboratories and in labs in the USAto learn how to build and operate a SR machine. Some have been hired in the meantime as SESAME staff. A new phase of the training programme was started when a Memorandum of Understanding was signed in December 2006 with IAEA providing funds for training of $ 750 000 for 4 years. Experts for beamlines are to be training as a next step. A number of special offers for training could be used. For example the National Synchrotron Radiation Research Centre (NSRRC) at Taiwan had funded three fellowships in 2005 and because of the satisfactory results they offered to continue the programme in 2006. Several workshops in different fields have been organized regularly to train future potential users.

In conclusion it may be stated that SESAME has passed the threshold of no-return, although some problems have still to be settled. The more the realisation of the project takes shape, the more interest and enthusiasm among potential users but also funding organisation and governments can be observed.

5. General Lessons learned from SESAME

Having been involved in SESAME for many years gave me the possibility to gain some general experience concerning science policy in the region and I dare to make some observations which might be useful for future activities.

a) The recognition that Science and Technology are important and necessary for any society and its welfare is one of the main conditions for the successful development of a country. To improve the awareness of this fact among political leaders is a conditio sine qua non. In the long-term satisfactory employment conditions, the manufacturing of technical products and the protection of the environment will not be possible without science-based technology. In Singapore about 65 % of all manufactured goods are high-tech products, in South Korea about 39 % wheras in the Gulf region it is less than 1 %.

Quite often priority is given to short-term problems (infrastructure, roads, water) and activities with short return (tourisms) are promoted. A mercantile mentality prevails in some countries.

b) Great achievements require outstanding persons in politics and science. The Role of individual political leaders is crucial (During the Golden Time of Science in Muslim Countries Khalifs, Princes and Vesirs found pleasure in doing or supporting sciences). In some Muslim countries such personalities exist, e.g. President Pervez Musharraf and minister Atta-ur Rahman, in Pakistan or HM Abdullah II and minister Kh.Toukan in Jordan and the very positive results are obvious.

For future scientific leaders a kind of career planning is necessary including the early transfer of responsibilities to younger scientists.

c) Developing countries need the best and most modern technologies. The most modern cars, power stations and excellent communication facilities for profession and entertainment. Only then will they be able to become full members of the international community respecting general conditions like environmental preservation.

d) The financial resources for R&D have to be increased considerably. The annual spending on R&D as a fraction of the GDP in the Islamic world is an order of magnitude below the global averages, in spite of the oil richness. No Arab country spends more than 0.2% of its GDP compared to a world average of 1.4% of GDP, industrialized countries spend about 2.5 % or even more than 3% of GDP (e.g. Denmark, France, Germany, Japan, Switzerland, the USA and

Finland)[†]. Several leaders in the Arab world are well aware of the precarious situation and make a big effort to improve it, but most of the times with little success. Many examples could be given. A recent example is the Summit meeting of the Arab League (22 nations) in Khartoum, Sudan, in March 2006 where Omar Hassan al-Bashir, president of Sudan, suggested that the increasing revenues from oil production should be used to fund science and technology development. So far no follow up has become known.

It would be sufficient to spend a very small amount (1 to 2 %) of the funds available for the general infrastructure for basic research, which should be considered as a long term investment.

e) Poor funding is not the only major reason for the bad situation of sciences. In many cases a clear strategy for research is missing. There is a lack of priority decisions for the research infrastructure and institutional support. Mechanisms for decision taking and priority setting have to be implemented. Evaluation procedures should be established and flexibility should be developed in redirecting funds if necessary.

f) It should be understood that scientific and technological know-how cannot be bought (even with a lot of oil money). Science is not a commodity that can be separated from the thought processes. For science and technology to flourish it needs a cultural base which can only be acquired by science education and doing research.

g) Sometimes there is a lack of cooperation inside individual countries. The creation of national networks should be encouraged.

h) International scientific collaboration is essential. It is the only way to define quality standards and achieve excellence. Many countries are accustomed to think in terms of national or at best in bilateral projects (e.g. with international organisations EU, IAEA, UNESCO, TWAS). This requires some experience in dealing with international bureaucracy.

i) Many scientists from the region have emigrated to industrialised countries. Some of them might be willing to return provided they get competitive research facilities. But at least their experience could be used for advice in many of the issues mentioned above.

[†] United Nations' Development Program and the Kuwait-based Arab Fund for Social and Economic Development, study 2003

In conclusion it is very gratifying to state that Pakistan is fulfilling many of the conditions mentioned above or is taking active steps in the directions suggested. The creation of the National Centre for Physics is an important element in such a development and my best wishes accompany it into a successful future.

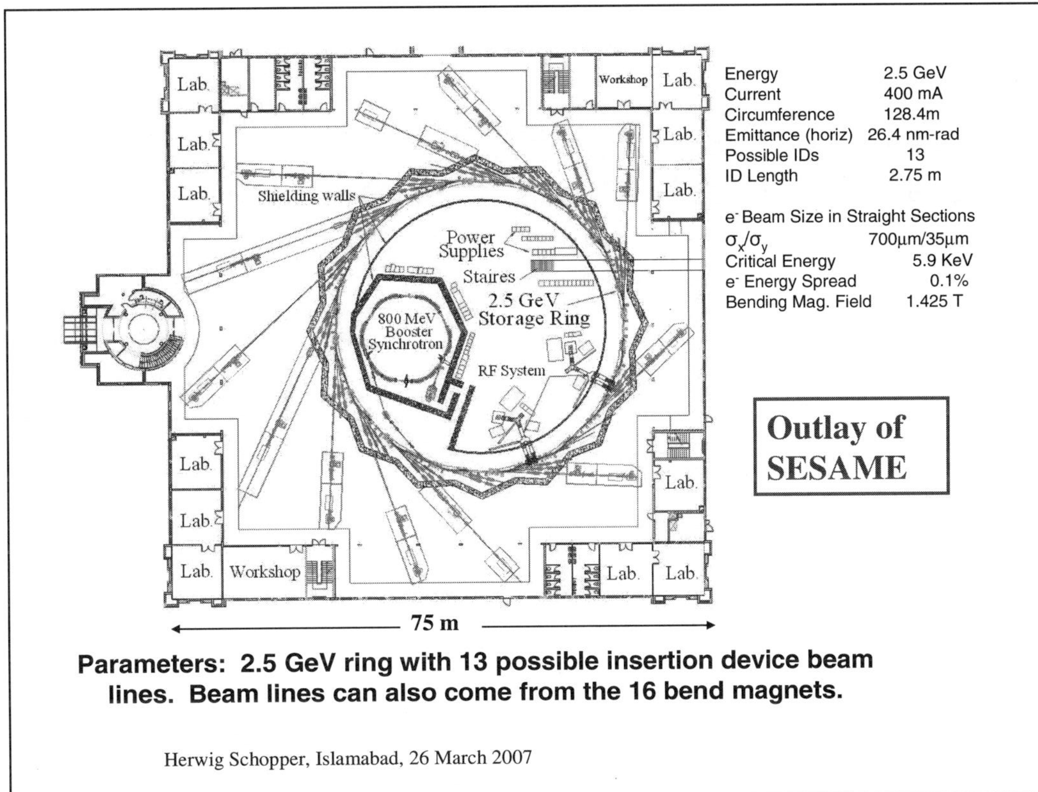

Fig.1. Layout of the SESAME Hall and main parameters of the facility.

Fig.2. The SESAME building housing the accelerator, laboratories, workshops and offices.

Phase One Beamlines

No.	Beamline	Energy Range	Source type	Science Area (traditional disciplines indicated)	Possible Champion & committee advisers
1	MAD Protein Crystallogphy	**4 – 14 keV**	**In-vacuum undulator (6mm gap, 1.5m)**	Structural Molecular Biology (Biomedical)	**Shoham, Salman, Yousef , Hasnain & Wakatsuki**
2	Photoemission/ Photoabsorption spectroscopy soft X rays	50-2000 eV	**Elliptically polarising undulator (λ_u=50 mm, k_{max}= 5.5, ID $_{length}$= 2.6 m)**	Atomic, Molecular & condensed matter physics (Physics, Chemistry, Materials)	**Hamdan, Shoaib Ahmed, (Zahid Hussain)**
3	SMALL/WIDE ANGLE X ray Scattering SAXS/WAXS	**8-12 keV**	**Undulator**	SMB & Material Sc (Biomedical, Physics, Chemistry, Materials engineers)	**Changizi, Semra Ide, Zehra Sayers, M. Al-Hussein, Irit Sagi**
4	X-ray Absorption Fine Structure/ Fluorescence Microprobe XAFS/XRF	**3-30 keV**	**2.0 Tesla MPWiggler**	SMB, MS, Environmental Sc., Archaeological	**Sagi, Mahmood, F. Afaneh, Rami Ali, Hallak (Hamdan & Hasnain)**
5	Powder Diffraction	3-25 keV	**2.4 Tesla MPWiggler**	MS, EnvironmentalSc., Archaeological	**Engin Ozdas, and Shoaib Akhtar**
6	IR Spectro-microscopy	0.01-1 eV	Large Aperture Bending magnet	SMB, MS, Environmental Sc., Archaeological	**El Bayyari & Sagi**
0	Soft X-Ray/UVU Spectroscopy	5 - 250 eV	Bending magnet	Start and learn	**SESAME** complete beam from LURE

Herwig Schopper, Islamabad, 26 March 2007

Fig.3. The first generation of planned beamlines covering the spectral range from infrared to X-rays.

The impact of synchroton light sources on science and society in developing countries

Herman Winick
SSRL/SLAC/Stanford University
E-mail: winick@slac.stanford.edu

At present there are more than 50 synchrotron radiation light source research facilities in operation in 19 countries around the world, with more in construction and design (see www.lightsources.org). Although most facilities are in technologically advanced countries, there is significant activity in the developing world. Twenty to thirty years ago several developing countries (Brazil, China, India, Korea, and Taiwan) began light source projects. These have now been completed and have been in operation for about 10 or more years. More recently a light source has started operation in Thailand (*http://www.nsrc.or.th/Eng/index.htm*). In this paper we describe the benefits that operating synchrotron light source facilities in developing countries have had.

Other projects are in progress. The SESAME project in Jordan (*www.sesame.org.jo*) is now constructing a 2.5 GeV third generation light source. A 3 GeV third generation light source is in design in Armenia (*http://www.candle.am/index.html*) and discussions are underway for a light source for southern Africa.

In the UNESCO-sponsored SESAME project eight countries (Bahrain, Cyprus, Egypt, Israel, Jordan, Palestinian Authority, Pakistan and Turkey) have joined to create a regional international synchrotron radiation research center which is now in construction in Jordan. The project started with equipment donated by Germany. In endorsing SESAME, the UNESCO Executive Board and General Assembly called it "a model for other regions" and "a quintessential UNESCO project for capacity building through science". Although the project is still in construction, significant benefits to the region have come from many workshops, schools, and Users' meetings in which hundreds of scientists from the region have met with scientists from around the world who have had experience using synchrotron radiation.

Most of these have focused on the many applications of synchrotron radiation to basic and applied research. In addition, fifty young engineers and scientists from the Middle East have been trained in accelerator technology in a 10 day school and eighteen of them have spent two years working with accelerator groups in European laboratories. Several of these are now on the accelerator staff of SESAME.

Pakistan has been a strong member of SESAME and is planning to construct a beam line on the storage ring. Several other countries have also announced their intention to build beam lines at SESAME. In addition, plans are being developed in Pakistan for a future 2.5-3 GeV third generation national light source, in anticipation of the growing community of users there.

Why is a synchrotron radiation facility relevant to a developing country? Before a light source is approved in a developing country a committee is usually formed to study the matter. The objectives and potential benefits identified by these committees include:

- Create a world-class interdisciplinary research laboratory
- Promote basic & applied research & technology
- Address regional biomedical & environmental issues/concerns
- Provide an environment for collaborations & individual development
- Train graduate students who will no longer have to go abroad
- Attract scientists working abroad to return
- Promote international scientific collaborations
- Promote development of high-tech industry (capacity building)

One of these, the UNESCO-sponsored SESAME Project in Jordan (*www.sesame.org.jo*), has an additional objective:

- Use scientific cooperation to promote peace & understanding between people from different traditions, religions, races, & political systems.

Below we summarize the impact that operational synchrotron radiation research facilities have had in several developing countries. The information presented was obtained from

- facility web sites
- presentations made at SRI2006 in Korea (see *http://sri2006.postech.ac.kr/*)

- inquiries in early 2006 to directors of operating facilities in developing countries

1. Laboratório Nacional de Luz Síncrotron (LNLS); Campinas, Brazil (ref. 1) *http://www.lnls.br/index.asp?idioma=2&opcaoesq*

A 1.35 GeV light source has been in operation in Campinas since1997. In 2004 there were more than one thousand external users. Demand for beam time exceeds available time by a factor of two. Twelve beam lines are in operation.

Since 2000 more than 29 PhD theses and 21 Masters Dissertations have been completed based on work done at LNLS. Almost all of the technical components of the facility were built in Brazil, largely by LNLS, which now has skills unique in Brazil in certain technical areas (ultra-high vacuum, magnets design and construction, controls, rf, etc). For example, as a result of the experience gained in constructing the storage ring, LNLS now supplies UHV equipment at cost to universities in Brazil. Of sixteen staff scientists four returned to Brazil after their postdoctoral research abroad.

LNLS plays an important role in the Latin-American scientific community (15% of LNLS users come from outside Brazil, including 9-10% from Argentina). Others come from Africa, Europe, and the USA.

LNLS has made significant contributions to industrial development in Brazil. One company got its start by building flanges for LNLS. In a permanent program to train industrial scientists, about 15 students work each year at LNLS for 6 months before taking industry jobs. Seven joint projects with industry are underway (e.g. catalysis). LNLS develops special instrumentation to meet the needs of industrial users. Seven LNLS staff are dedicated to service to industry.

LNLS has introduced the concept of a national research facility, equipped with world-class equipment and open to all.

2. Beijing Synchrotron Radiation Facility (BSRF); Beijing, China: (ref. 2) *http://www.ihep.ac.cn/bsrf/english/main/main.htm*

A 2.5 GeV light source has been in operation since the early 1990's. The storage ring is primarily used as a high energy physics colliding beam facility, and partly used for synchrotron radiation research. A major upgrade program for both high energy physics and synchrotron radiation has recently been completed. This included a new experimental hall and a new wiggler beam line.

In the period from 1999 to 2006 about 100 students completed a PhD or Masters Degree and fifteen completed postdoctoral work. Three former students returned after periods abroad as postdocs or staff members at other institutions.

Before the shutdown for the upgrade 13 beam lines and 13 experimental stations were in operation (6 wiggler beam lines and 7 bending magnet beam lines) serving 1087 users from 21 domestic cities. Users came from 86 (including 6 foreign) Universities, Institutes and companies (catalysis, micro-electronics and others).

3. National Synchrotron Radiation Laboratory (NSRL); Hefei, China (ref. 3) *http://en.nsrl.ustc.edu.cn/*

An 0.8 GeV light source has been in operation since 1991. A major Phase II upgrade was carried out from 1999 to 2004. NSRL now operates 14 photon beamlines and experimental stations, three of them from a 6 Tesla superconducting wave length shifter (SLS) and one from a 29-period permanent magnet undulator. The facility operates for 4,800 hours per year and serves 400 users.

58 PhD and 53 Masters Degrees have been produced by the University of Science and Technology of China alone. This now continues at the yearly rate of 20 PhD and 20 Masters Degrees per year. A total of about 150 graduate students are now involved. Several senior scientists have returned to NSRL from abroad, including Chen Gao from LBNL, now Vice-Director of NSRL.

4. Pohang Accelerator Laboratory (PAL); Pohang, Korea (ref. 4) *http://pal.postech.ac.kr/eng/index.html*

A 2.5 GeV light source has been in operation since 1994. Twenty seven beam lines, inclduding six from insertion devices, are in operation, serving 1500-2000 users each year. About 500 papers are published each year based on work at the light source.

Ten scientists returned to Korea to join the laboratory. Four became directors. Thirty professors who returned to the Pohang University of Science and Technology (Postech) are now users of the light source. Many others returned to become faculty at other Korean Universities and are now as users at PLS. At least 40 PhDs have already been completed at Postech using the light source. About 100 PhDs have been completed at other universities in Korea using the light source.

Many parts of PLS have been built in Korea (magnets, power supplies, vacuum chambers, ion pumps, klystrons, modulators, beam position monitors, etc.). The 2.5 GeV linac injector was designed and constructed in collaboration with China, although China & Korea did not have diplomatic relations until 2 years after this collaboration started. Collaborations with other countries included France, Italy, Japan, the UK, and the US.

Several new companies were formed as a direct result of the light source project. These include Geumryong (magnets), VMT (ion pumps), a controls company, and an alignment company. Samsung & LG studied defects in LSI chips & LCDs for cellular phones.

A comment by a former director: "As a result of the success of the PLS project, ordinary Koreans and the Korean government trust Korean scientists since they spent a large amount of money and achieved, and even exceeded, project goals. Initially there was some doubt that this could be done."

5. National Synchrotron Radiation Research Center (NSRRC) Hsinchu,Taiwan (ref. 5) *http://www.nsrrc.org.tw/*

A 1.5 GeV light source has been in operation since 1993. A total of 28 beamlines, 45 end stations, and 8 insertion device magnets are in operation. These serve about 1000 users on 660 experimental proposals in 2005. Also two hard x-ray beam lines (one from a bending magnet and one from an undulator) have been constructed at Spring 8 in Japan for use by scientists from Taiwan. These have been in operation since 2001.

So far more than 140 students have received the PhD based on work done at the light source. More than 25 overseas Taiwanese scientists returned to NSRRC, universities, and national research institutes in Taiwan due at least in part to the light source.

There has been significant benefit to companies in Taiwan which have collaborated with the light source to develop capabilities to design, fabricate, and measure many technical components. This includes magnets, insertion devices (including superconducting devices), ultra-high vacuum chambers and other components, survey and alignment technology, beam line components, x-ray energy analyzers, cryogenics technology, radio frequency components, digital electronics, and brazing technology.

As part of the international synchrotron radiation community, the light source has had many interactions with other facilities, including Spring 8, ANKA, ALS, SSRL, CHESS, the Australian light source, and SESAME. The Taiwan light source has provided full support for three fellowships for scientists from the Middle East who worked at NSRRC for one year. Three more have been offered.

A new 3-3.3 GeV third generation light source is proposed at the same site. This proposal is currently under evaluation by the National Science Council of Taiwan.

6. National Synchrotron Research Center (NSRC) Nakhon Ratchasima, Thailand (ref. 6) *http://www.nsrc.or.th/Eng/index.htm*

A 1 GeV light source has been in operation since 2001 using equipment donated by Japan. It has recently been upgraded to 1.2 GeV. Three beam lines are now in operation. A 6.4 T wavelength shifter is now being installed. This will provide a spectrum with a critical photon energy of 6.15 keV and useful intensity up to 20-25 keV. Seven students have completed PhDs and two have completed Masters Degrees. Presently, ten PhDs and four Masters Degrees are in progress.

References

1. Jose Brum; personal communication
2. Ding Chang Xian; personal communication
3. Zupin Liu; personal communication
4. In Soo Ko; personal communication
5. Keng Liang; personal communication
6. Helmut Wiedemann; personal communication

SECTION II

QUANTUM PHYSICS AND QUANTUM INFORMATION

Discrimination of quantum states with selected applications

János A. Bergou

Department of Physics and Astronomy,
Hunter College of the City University of New York,
695 Park Avenue, New York, NY 10021, USA

In quantum information processing and quantum computing protocols information is encoded in the state of a quantum system. After processing the information it has to be read out what is equivalent to determining the final state of the system. When the possible final states are not orthogonal this is a highly nontrivial task that constitutes the general area of what is known as quantum state discrimination. It consists in finding measurement schemes that, according to some figure of merit, will determine the state of the system. Optimized measurement schemes often lead to generalized measurements (Positive Operator Valued Measures, POVMs). In this tutorial review we illustrate the power of the POVM concept on examples relevant to applications in quantum cryptography. In order to keep the flow of the presentation we give a brief introduction to the theory of generalized measurements in an appendix.

1. Introduction

In quantum information and quantum computing the carrier of information is a quantum system and information is encoded in its state.[1] The state, however, is not an observable in quantum mechanics[2] and, thus, a fundamental problem arises: after processing the information - i.e. after the desired transformation is performed on the input state by the quantum processor - the information has to be read out or, in other words, the state of the system has to be determined. When the set of possible target states is known and the states in the set are mutually orthogonal, this is a relatively simple task. One simply has to set up detectors along these orthogonal directions and a click in one of the detectors will unambiguously identify the state of the system. However, when the possible target states are not orthogonal they cannot be discriminated perfectly, and optimum discrimination with respect to some appropriately chosen criteria is far from being trivial even if the set of the possible nonorthogonal states is known. The problem of discriminating among nonorthogonal states is ubiquitous in quantum information and quantum computing, underlying many of the communication (cryptography) and computing (probabilistic algorithms) schemes that have been suggested so far. It is, in general, a measurement optimization problem (more about this aspect can be found in recent reviews[3,4]). It is the purpose of this contribution to introduce the various theoretical and experimental tools that have been developed for discriminating among nonorthogonal quantum states. Interestingly, the problem of discriminating among nonorthogonal quantum states was introduced by the seminal works of Helstrom[5] and Holevo[6] long before the term Quantum Information Theory was even coined. Stimulated by the rapid developments in quantum information theory of the 90's, the question of how to discriminate between nonorthogonal quantum states *in an optimum way* has gained new momentum and state discrimination quickly became an integral part of the quantum information theory toolbox. In particular, the suggestion to use nonorthogonal quantum states for communication in secure quantum cryptographic protocols, most notably in the quantum key distribution (QKD) scheme based on the two-state procedure as developed by Bennett[7] (henceforth referred to as the B92 protocol), has given major impetus to this area.

In order to devise an optimum state-discriminating measurement, strategies have been developed with respect to various criteria. Often, these optimized strategies involve generalized measurements (Positive Operator Valued Measures, POVMs). In order to keep the flow of the presentation, in the Appendix A we give a brief overview of the standard quantum measurements, in Appendix B we introduce the concept of Positive Operator Valued Measures (POVMs) and in Appendix C we discuss their experimental implementation via generalized measurements based on Neumark's theorem. As for illustrative applications of the theory, in Sec. 2 we discuss the two main strategies for discriminating between different states of a quantum system. The two methods are optimum unambiguous discrimination of the states, and state discrimination with minimum error. They will be outlined in detail in the subsections 2.1 and 2.2. As an example for the application of these two main discrimination strategies, optimal figures of merit for the B92 protocol will be reviewed in Sec. 3. Section 4 is devoted to the recently emerging sub-field of discriminating among mixed states. We conclude with a brief outlook in Sec. 5.

2. State discrimination strategies

As examples of a measurement optimization task we will consider two schemes for the optimal discrimination of quantum states. The first is unambiguous discrimination and the second is discrimination with minimum error. We will see that the optimum measurement for the first strategy is a POVM while the optimum measurement for the second is a standard von Neumann measurement. The two main discrimination strategies evolved rather differently from the very beginning. Unambiguous discrimination started with pure states and only very recently was it extended to discriminating among mixed quantum states. Minimum-error discrimination addressed the problem of discriminating between two mixed quantum states from the very beginning and the result for two pure states follows as a special case. The two strategies are, in a sense, complementary to each other. Unambiguous discrimination is relatively straightforward to generalize for more than two states, at least in principle, but it is difficult to treat mixed states. The error-minimizing approach, initially developed for two mixed states, is hard to generalize for more than two states.

2.1. *Unambiguous discrimination of two pure states*

Unambiguous discrimination is concerned with the following problem. An ensemble of quantum systems is prepared so that each individual system is prepared in one of two known states, $|\psi_1\rangle$ or $|\psi_2\rangle$ with probability η_1 or η_2 (such that $\eta_1 + \eta_2 = 1$), respectively. The preparation probabilities are called a priori probabilities or, simply, priors. The states are, in general, not orthogonal, $\langle\psi_1|\psi_2\rangle \neq 0$ but linearly independent. The preparer, Alice, then draws a system at random from this ensemble and hands it over to an observer, called Bob, whose task is to determine which one of the two states he is given. The observer also knows how the ensemble was prepared, i.e. has full knowledge of the two possible states and their priors but does not know the actual state that was drawn. All he can do is to perform a single measurement or perhaps a POVM on the individual system he receives.

In the unambiguous discrimination strategy the observer is not allowed to make an error, i.e. he is not permitted to conclude that he was given one state when actually he was given the other. First we show that this can not be done with 100% probability of success. To this end, let us assume the contrary and assume we have two detection operators, Π_1 and Π_2, that together span the Hilbert space of the two states,

$$\Pi_1 + \Pi_2 = I \ . \tag{1}$$

For unambiguous detection we also require that

$$\Pi_1|\psi_2\rangle = 0 \ ,$$
$$\Pi_2|\psi_1\rangle = 0 \ , \tag{2}$$

so that the first detector never clicks for the second state and vice versa, and we can identify the detector clicks with one of the states unambiguously. The probability of successfully identifying the first state if it is given is $p_1 = \langle\psi_1|\Pi_1|\psi_1\rangle$ and the probability of successfully identifying the second state if it is given is $p_2 = \langle\psi_2|\Pi_2|\psi_2\rangle$. Multiplying (1) with $\langle\psi_1|$ from the left and $|\psi_1\rangle$ from the right and taking into account (2), gives $p_1 = 1$ and, similarly, we obtain $p_2 = 1$, and it appears as though we could have perfect unambiguous discrimination. However, multiplying (1) with $\langle\psi_1|$ from the left and $|\psi_2\rangle$ from the right and taking into account (2) again, gives $0 = \langle\psi_1|\psi_2\rangle$ which can be satisfied for orthogonal states only. In fact, we have just proved that perfect discrimination of nonorthogonal quantum states is not possible.

Equation (1) allows two alternatives only, it assumes that we can have two operators that unambiguously identify the two states all the time. Since this is impossible, we are forced to modify this equation and have to allow for one other alternative. We introduce a third POVM element, Π_0, such that Eq. (2) is still satisfied but (1) is modified to

$$\Pi_1 + \Pi_2 + \Pi_0 = I \ . \tag{3}$$

The first and second POVM elements will continue to unambiguously identify the first and second state, respectively. However, Π_0 can click for both states and, thus, this POVM element corresponds to an inconclusive detection result. It should be emphasized that this outcome is not an error, we will never identify the first state with the second and vice versa, we simply will not make any conclusion in this case. We can now introduce success and failure probabilities in such a way that $\langle\psi_1|\Pi_1|\psi_1\rangle = p_1$ is the probability of successfully identifying $|\psi_1\rangle$, and $\langle\psi_1|\Pi_0|\psi_1\rangle = q_1$ is the probability of failing to identify $|\psi_1\rangle$, (and similarly for $|\psi_2\rangle$). For unambiguous discrimination we have $\langle\psi_2|\Pi_1|\psi_2\rangle = \langle\psi_1|\Pi_2|\psi_1\rangle = 0$ from (2). Using this, we obtain from Eq. (3) $p_1 + q_1 = p_2 + q_2 = 1$. This means that if we allow inconclusive detection results to occur with a certain probability then in the remaining cases the observer can conclusively determine the state of the individual system.

It is rather easy to see that a simple von Neumann measurement can accomplish this task. Let us denote the Hilbert space of the two given states by $\mathcal{H}$ and introduce the projector P_1 for $|\psi_1\rangle$ and $\bar{P}_1$ for the orthogonal subspace, such that $P_1 + \bar{P}_1 = I$, the identity in $\mathcal{H}$. Then we know for sure that $|\psi_2\rangle$ was prepared if in the measurement of $\{P_1, \bar{P}_1\}$ a click in the $\bar{P}_1$ detector occurs. A similar conclusion for $|\psi_1\rangle$ can be reached with the roles of $|\psi_1\rangle$ and $|\psi_2\rangle$ reversed. Of course, when a click along P_1 (or P_2) occurs then we learn nothing about which state was prepared, this outcome thus corresponding to the inconclusive result. In the von Neumann set-ups one of the alternatives is missing. We either identify one state or we get an inconclusive result but we miss the other state completely. This scenario is actually allowed by (3).

We now turn our attention to the determination of the optimum measurement strategy for unambiguous discrimination. It is the strategy, or measurement set-up, for which the average failure probability is a minimum (or, equivalently, the average success probability is a maximum). We want to determine the operators in (3) explicitly. If we introduce $|\psi_j^\perp\rangle$ as the vector orthogonal to $|\psi_j\rangle$ $(j = 1, 2)$ then the condition of unambiguous detection, Eq. (2), mandates the choices

$$\Pi_1 = c_1 |\psi_2^\perp\rangle\langle\psi_2^\perp| \,, \tag{4}$$

and

$$\Pi_2 = c_2 |\psi_1^\perp\rangle\langle\psi_1^\perp| \,. \tag{5}$$

Here c_1 and c_2 are positive coefficients to be determined from the condition of optimum.

Inserting these expressions in the definition of p_1 and p_2 gives $c_1 = p_1/|\langle\psi_1|\psi_2^\perp\rangle|^2$ and a similar expression for c_2. Finally, introducing $\cos\Theta = |\langle\psi_1|\psi_2\rangle|$ and $\sin\Theta = |\langle\psi_1|\psi_2^\perp\rangle|$, we can write the detection operators as

$$\begin{aligned} \Pi_1 &= \frac{p_1}{\sin^2\Theta} |\psi_2^\perp\rangle\langle\psi_2^\perp| \,, \\ \Pi_2 &= \frac{p_2}{\sin^2\Theta} |\psi_1^\perp\rangle\langle\psi_1^\perp| \,. \end{aligned} \tag{6}$$

Now, Π_1 and Π_2 are positive semi-definite operators by construction. However, there is one additional condition for the existence of the POVM which is the positivity of the inconclusive detection operator,

$$\Pi_0 = I - \Pi_1 - \Pi_2 \,. \tag{7}$$

This is a simple 2 by 2 matrix in $\mathcal{H}$ and the corresponding eigenvalue problem can be solved analytically. Non-negativity of the eigenvalues leads, after some tedious but straightforward algebra, to the condition

$$q_1 q_2 \geq |\langle\psi_1|\psi_2\rangle|^2 \,, \tag{8}$$

where $q_1 = 1 - p_1$ and $q_2 = 1 - p_2$ are the failure probabilities for the corresponding input states.

Eq. (8) represents the constraint imposed by the positivity requirement on the optimum detection operators. The task we set out to solve can now be formulated as follows. Let

$$Q = \eta_1 q_1 + \eta_2 q_2 \tag{9}$$

denote the average failure probability for unambiguous discrimination. We want to minimize this failure probability subject to the constraint, Eq. (8). Due to the relation, $P = \eta_1 p_1 + \eta_2 p_2 = 1 - Q$, the minimum of Q also gives us the maximum probability of success. Clearly, for optimum the product $q_1 q_2$ should be at its minimum allowed by (8), and we can then express q_2 with the help of q_1 as $q_2 = \cos^2\Theta/q_1$. Inserting this expression in (9) yields

$$Q = \eta_1 q_1 + \eta_2 \frac{\cos^2\Theta}{q_1} , \tag{10}$$

where q_1 can now be regarded as the independent parameter of the problem. Optimization of Q with respect to q_1 gives $q_1^{POVM} = \sqrt{\eta_2/\eta_1}\cos\Theta$ and $q_2^{POVM} = \sqrt{\eta_1/\eta_2}\cos\Theta$. Finally, substituting these optimal values into Eq. (9) gives the optimum failure probability,

$$Q^{POVM} = 2\sqrt{\eta_1\eta_2}\cos\Theta . \tag{11}$$

Let us next see how this result compares to the average failure probabilities of the two possible unambiguously discriminating von Neumann measurements that were described at the beginning of this section. The average failure probability for the first von Neumann measurement, with its failure direction along $|\psi_1\rangle$, can be written by simple inspection as

$$Q_1 = \eta_1 + \eta_2|\langle\psi_1|\psi_2\rangle|^2 , \tag{12}$$

since $|\psi_1\rangle$ gives a click with probability 1 in this direction but it is only prepared with probability η_1 and $|\psi_2\rangle$ gives a click with probability $|\langle\psi_1|\psi_2\rangle|^2$ but it is only prepared with probability η_2.

By entirely similar reasoning, the average failure probability for the second von Neumann measurement, with its failure direction along $|\psi_2\rangle$, is given by

$$Q_2 = \eta_1|\langle\psi_1|\psi_2\rangle|^2 + \eta_2 . \tag{13}$$

What we can observe is that Q_1 and Q_2 are given as the arithmetic mean of two terms and Q^{POVM} is the geometric mean of the same two terms for either case. So, one would be tempted to say that the POVM performs better always. This, however, is not quite the case, it does so only when it exists. The obvious condition for the POVM solution to exist is that both $q_1^{POVM} \leq 1$ and $q_2^{POVM} \leq 1$. Using $\eta_2 = 1 - \eta_1$, a little algebra tells us that the POVM exists in the range $\cos^2\Theta/(1+\cos^2\Theta) \leq \eta_1 \leq 1/(1+\cos^2\Theta)$. If η_1 is smaller than the lower boundary, the POVM goes over to the first von Neumann measurement and if η_1 exceeds the upper boundary the POVM goes over to the second von Neumann measurement. This can be easily seen from Eqs. (6) and (7) since $p_1 = 1 - q_1 = 0$ for $q_1 = 1$ and Π_0 becomes a projection along $|\psi_1\rangle$ (and correspondingly for $p_2 = 0$).

These findings can be summarized as follows. The optimal failure probability, Q^{opt}, is given as

$$Q^{opt} = \begin{cases} Q^{POVM} & \text{if } \frac{\cos^2\Theta}{1+\cos^2\Theta} \leq \eta_1 \leq \frac{1}{1+\cos^2\Theta} \,, \\ Q_1 & \text{if } \eta_1 < \frac{\cos^2\Theta}{1+\cos^2\Theta} \,, \\ Q_2 & \text{if } \frac{1}{1+\cos^2\Theta} < \eta_1 \,. \end{cases} \tag{14}$$

The optimum detection operators are given by

$$\Pi_1 = \frac{1 - q_1^{opt}}{\sin^2\Theta} |\psi_2^\perp\rangle\langle\psi_2^\perp| \,,$$
$$\Pi_2 = \frac{1 - q_2^{opt}}{\sin^2\Theta} |\psi_1^\perp\rangle\langle\psi_2^\perp| \,. \tag{15}$$

These expressions show explicitly that $\Pi_1 = 0$ and Π_2 is the projector $|\psi_1^\perp\rangle\langle\psi_1^\perp|$ when $q_1^{opt} = 1$ and $q_2^{opt} = \cos^2\Theta$, i.e. the POVM goes over smoothly into a projective measurement at the lower boundary and, similarly, into the other von Neumann projective measurement at the upper boundary.

The above result is very satisfying from a physical point of view. The POVM delivers a lower failure probability in its entire range of existence than either of the two von Neumann measurements. At the boundaries of this range it merges smoothly with the one von Neumann measurement that has a lower failure probability at that point. Outside this range the state preparation is dominated by one of the states and the optimal measurement becomes a von Neumann projective measurement, using the state that is prepared less frequently as its failure direction. It should be noted that unambiguous discrimination was first suggested by Ivanovic[8], the POVM leading to Eq. (11) when $\eta_1 = \eta_2 = 1/2$ was found by Dieks[9] and Peres[10] proved its optimality, therefore it is called the IDP limit in this case. For arbitrary prior probabilities Jaeger and Shimony[11] derived Eq. (14).

To close this subsection we note the optimal unambiguous discrimination of more than two states is still a partially open problem. We have obtained analytical solutions for some special cases of discriminating among three pure states[12] that have subsequently been verified experimentally[13].

2.2. *Minimum error discrimination of two quantum states*

In the previous section we have required that, whenever a definite answer is returned after a measurement on the system, the result should be unambiguous, at the expense of allowing inconclusive outcomes to occur. For many applications in quantum communication, however, one wants to have conclusive results only. This means that errors are unavoidable when the states are non-orthogonal. Based on the outcome of the measurement, in each single case then a guess has to be made as to what the state of the quantum system was. In the optimal strategy we want to minimize the probability of making a wrong guess, hence this procedure is known as *minimum error* discrimination. The problem is to find the optimum measurement that minimizes the probability of errors.

Let us state the optimization problem a little more precisely. In the most general case, we want to distinguish, with minimum probability of error, among N given states of a quantum system (where $N \geq 2$). The states are given by the density operators ρ_j ($j = 1, 2, \ldots, N$) and the jth state occurs with the given a priori probability η_j, such that $\sum_{j=1}^{N} \eta_j = 1$. The measurement can be formally described with the help of a POVM, where the POVM elements, Π_j, correspond to the possible measurement outcomes. They are defined in such a way that $\text{Tr}(\rho\Pi_j)$ is the probability to infer the state of the system to be ρ_j if it has been prepared in a state ρ. Since the probability is a real non-negative number, the detection operators once again have to be positive-semidefinite. In the error-minimizing measurement scheme the measurement is required to be exhaustive and conclusive in the sense that in each single case one of the N possible states is identified with certainty and inconclusive results do not occur. This leads to the requirement

$$\sum_{j=1}^{N} \Pi_j = I_{D_S}, \tag{16}$$

where I_{D_S} denotes the identity operator in the D_S-dimensional physical state space of the quantum system. The overall probability P_{err} to make an erroneous guess for any of the incoming states is then given by

$$P_{\text{err}} = 1 - P_{\text{corr}} = 1 - \sum_{j=1}^{N} \eta_j \text{Tr}(\rho_j \Pi_j) \tag{17}$$

with $\sum_j \eta_j = 1$. Here we introduced the probability P_{corr} that the guess is correct. In order to find the minimum-error measurement strategy, one has to determine the POVM that minimizes the value of P_{err} under the constraint given by Eq. (16). By inserting these optimum detection operators into Eq. (17), the minimum error probability $P_{\text{err}}^{\text{min}} \equiv P_E$ is determined. The explicit solution to the error-minimizing problem is not trivial and analytical expressions have been derived only for a few special cases.

For the case that only two states are given, either pure or mixed, the minimum error probability, P_E, was derived in the mid 1970s by Helstrom in the framework of quantum detection and estimation theory. We find it more instructive to start by analyzing the two-state minimum-error measurement with the help of an alternative method that allows us to gain immediate insight into the structure of the optimum detection operators, without applying variational techniques. Starting from Eq. (17) and making use of the relations $\eta_1 + \eta_2 = 1$ and $\Pi_1 + \Pi_2 = I_{D_S}$ that have to be fulfilled by the a priori probabilities and the detection operators, respectively, we see that the total probability to get an erroneous result in the measurement is given by

$$P_{\text{err}} = 1 - \sum_{j=1}^{2} \eta_j \text{Tr}(\rho_j \Pi_j) = \eta_1 \text{Tr}(\rho_1 \Pi_2) + \eta_2 \text{Tr}(\rho_2 \Pi_1). \tag{18}$$

This can be alternatively expressed as

$$P_{\text{err}} = \eta_1 + \text{Tr}(\Lambda \Pi_1) = \eta_2 - \text{Tr}(\Lambda \Pi_2), \tag{19}$$

where we introduced the Hermitian operator

$$\Lambda = \eta_2 \rho_2 - \eta_1 \rho_1 = \sum_{k=1}^{D_S} \lambda_k |\phi_k\rangle\langle\phi_k|. \tag{20}$$

Here the states $|\phi_k\rangle$ denote the orthonormal eigenstates belonging to the eigenvalues λ_k of the operator Λ. The eigenvalues are real, and without loss of generality we can number them in such a way that

$$\begin{aligned} \lambda_k &< 0 \qquad \text{for} \qquad 1 \leq k < k_0, \\ \lambda_k &> 0 \qquad \text{for} \qquad k_0 \leq k \leq D, \\ \lambda_k &= 0 \qquad \text{for} \qquad D < k \leq D_S. \end{aligned} \tag{21}$$

By using the spectral decomposition of Λ, we get the representations

$$P_{\text{err}} = \eta_1 + \sum_{k=1}^{D_S} \lambda_k \langle\phi_k|\Pi_1|\phi_k\rangle = \eta_2 - \sum_{k=1}^{D_S} \lambda_k \langle\phi_k|\Pi_2|\phi_k\rangle. \tag{22}$$

Our optimization task now consists in determining the specific operators Π_1, or Π_2, respectively, that minimize the right-hand side of Eq. (22) under the constraint that

$$0 \leq \langle\phi_k|\Pi_j|\phi_k\rangle \leq 1 \qquad (j = 1, 2) \tag{23}$$

for all eigenstates $|\phi_k\rangle$. The latter requirement is due to the fact that $\mathrm{Tr}(\rho\Pi_j)$ denotes a probability for any ρ. From this constraint and from Eq. (22) it immediately follows that the smallest possible error probability, $P_{\text{err}}^{\text{min}} \equiv P_E$, is achieved when the detection operators are chosen in such a way that the equations $\langle\phi_k|\Pi_1|\phi_k\rangle = 1$ and $\langle\phi_k|\Pi_2|\phi_k\rangle = 0$ are fulfilled for eigenstates belonging to negative eigenvalues, while eigenstates corresponding to positive eigenvalues obey the equations $\langle\phi_k|\Pi_1|\phi_k\rangle = 0$ and $\langle\phi_k|\Pi_2|\phi_k\rangle = 1$. Hence the optimum detection operators can be written as

$$\Pi_1 = \sum_{k=1}^{k_0-1} |\phi_k\rangle\langle\phi_k|, \qquad \Pi_2 = \sum_{k=k_0}^{D_S} |\phi_k\rangle\langle\phi_k|, \tag{24}$$

where the expression for Π_2 has been supplemented by projection operators onto eigenstates belonging to the eigenvalue $\lambda_k = 0$, in such a way that $\Pi_1 + \Pi_2 = I_{D_S}$. Obviously, provided that there are positive as well as negative eigenvalues in the spectral decomposition of Λ, the minimum-error measurement for discriminating two quantum states is a von Neumann measurement that consists in performing projections onto the two orthogonal subspaces spanned by the set of states $\{|\phi_1\rangle, \ldots, |\phi_{k_0-1}\rangle\}$, on the one hand, and $\{|\phi_{k_0}\rangle, \ldots, |\phi_{D_S}\rangle\}$, on the other hand. An interesting special case arises when negative eigenvalues do not exist. In this case it follows that $\Pi_1 = 0$ and $\Pi_2 = I_{D_S}$ which means that the minimum error probability can be achieved by always guessing the quantum system to be in the state ρ_2, without performing any measurement at all. Similar considerations hold true in the absence of positive eigenvalues so a measurement does not always aid minimum-error discrimination. By inserting the optimum detection operators into Eq. (19) the minimum error probability is found to be

$$P_E = \eta_1 - \sum_{k=1}^{k_0-1} |\lambda_k| = \eta_2 - \sum_{k=k_0}^{D} |\lambda_k|. \tag{25}$$

Taking the sum of these two alternative representations and using $\eta_1 + \eta_2 = 1$, we arrive at

$$P_E = \frac{1}{2}\left(1 - \sum_k |\lambda_k|\right) = \frac{1}{2}\left(1 - \mathrm{Tr}|\Lambda|\right), \tag{26}$$

where $|\Lambda| = \sqrt{\Lambda^\dagger\Lambda}$. Together with Eq. (17) this immediately yields the well-known Helstrom formula for the minimum error probability in discriminating ρ_1 and ρ_2,

$$P_E = \frac{1}{2}\left(1 - \mathrm{Tr}|\eta_2\rho_2 - \eta_1\rho_1|\right) = \frac{1}{2}\left(1 - \||\eta_2\rho_2 - \eta_1\rho_1\|\right). \tag{27}$$

In the special case that the states to be distinguished are the pure states $|\psi_1\rangle$ and $|\psi_2\rangle$, this expression reduces to

$$P_E = \frac{1}{2}\left(1 - \sqrt{1 - 4\eta_1\eta_2|\langle\psi_1|\psi_2\rangle|^2}\right). \tag{28}$$

This expression, which is the one found in textbooks, can be cast to the equivalent form,

$$P_E = \eta_{min}\left(1 - \frac{2\eta_{max}(1 - |\langle\psi_1|\psi_2\rangle|^2)}{\eta_{max} - \eta_{min} + \sqrt{1 - 4\eta_{min}\eta_{max}|\langle\psi_1|\psi_2\rangle|^2}}\right), \tag{29}$$

where η_{min} (η_{max}) is the smaller (greater) of the prior probabilities, η_1 and η_2. This form lends itself to a transparent interpretation. The first factor on the right-hand-side is what we would get if we always guessed the state that is prepared more often, without any measurement at all. Thus, the factor multiplying η_{min} is the result of the optimized measurement.

The set-up of the detectors that achieve the optimum error probabilities is particularly simple for the case of equal a priori probabilities. Two orthogonal detectors, placed symmetrically around the two pure states, will do the task. The simplicity is particularly striking when one compares this set-up to the corresponding POVM set-up for optimal unambiguous discrimination.

As we mentioned already in the introduction, the result Eq. (27) was obtained independently by Helstrom[5] and Holevo[6] and is commonly referred to as the Helstrom bound. The very pedagogical derivation presented was given by Herzog[14].

Finally, we present an interesting relation, without proof, that is always satisfied between the minimum-error probability of the minimum-error detection and the optimal failure probability of unambiguous detection[14]. It reads as

$$P_E \leq \frac{1}{2}Q^{opt}. \tag{30}$$

This result hold for pure as well as mixed states, so it paves the road to mixed state discrimination to be reviewed in Sec. 4. It means that for two arbitrary states (mixed or pure), prepared with arbitrary a priori probabilities, the smallest possible failure probability in unambiguous discrimination is at least twice as large as the smallest probability of errors in minimum-error discrimination of the same states.

Fig. 1 summarizes the results of this Section. In it we display the failure probabilities, Q_1, Q_2, and Q^{POVM} from Eq. (14) as well as the minimum error probability P_E from Eq. (29) vs. η_1 for a fixed value of the overlap, $\cos^2\Theta$.

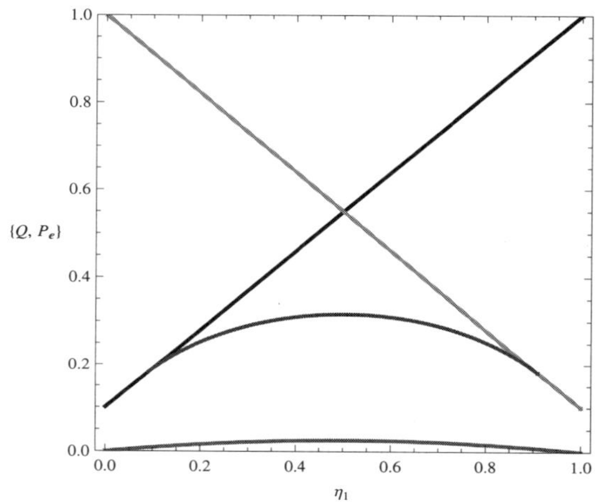

Fig. 1. Failure probability, Q, and minimum error probability, P_E, vs. the prior probability, η_1. Blue line: Q_1, green line: Q_2, red line: Q^{POVM}, magenta line: P_E. For the figure we used the following representative value: $|\langle\psi_1|\psi_2\rangle|^2 = 0.1$. For this the optimal failure probability, Q_{opt} is given by Q_1 for $0 < \eta_1 < 0.09$, by Q^{POVM} for $0.09 \leq \eta_1 \leq 0.9$ and by Q_2 for $0.9 < \eta_1$. Note that the inequality in Eq. (30) is always obeyed.

3. Example: The B92 quantum key distribution protocol

As an application of the material developed in the preceding Section we will now analyze the B92 protocol for quantum key distribution (QKD). QKD is the crucial ingredient of most quantum cryptographic protocols and in the B92 proposal all of the concepts of the previous Section come together in a particularly clean and instructive form.

In cryptography the sender is often called Alice and the receiver Bob. We will use this nomenclature in what follows. The basic task in cryptography can then be formulated as this. Alice wants to send a message to Bob, and keep it secret from everybody else. To accomplish this, she uses a code to encrypt the message. A possible code is to shift each letter of the message by a different amount. In this case Alice and Bob must share a sequence of numbers telling them the shifts. This is the key. If the key is random and is used only once the code is unbreakable. The question is, then, how to generate a secret key?

In 1992 Bennett[7] proposed using the unambiguous discrimination of two nonorthogonal states as the basis of a form of quantum cryptography. Quantum cryptography is a method of generating a secure shared key by quantum mechanical means that is discarded after being used only once. So, it is the quantum version of the one-time pad cipher and here is how it works.

(1) Alice sends a system that is prepared with equal probability either in the state $|\psi_0\rangle = |0\rangle$, which corresponds to the logical 0, or in the state $|\psi_1\rangle = \frac{1}{\sqrt{2}}(|0\rangle + |1\rangle)$, which corresponds to the logical 1.
(2) Bob applies optimum unambiguous state discrimination strategy to the state he receives. Using Eq. (11), the success probability for Bob's measurement is $P = 1 - Q^{POVM} = 1 - \frac{1}{\sqrt{2}}$.
(3) Bob tells Alice, over a public classical channel, whether the discrimination succeeded or failed. They keep the bit if the discrimination was successful, throw it away if it failed.

After repeating this procedure many times Alice and Bob share a sequence of 0's and 1's that they can use for a key.

Why is this procedure secure? Suppose an eavesdropper, called Eve, has intercepted the particle. She cannot determine whether it is in $|\psi_0\rangle$ or $|\psi_1\rangle$. One thing she can do is to apply the optimum unambiguous state discrimination procedure. Then she will fail with a probability of $\frac{1}{\sqrt{2}} \approx 71\%$. When she does, she has no idea what state was sent, so she must guess which one to send to Bob. Since the two states are prepared with equal probability, Eve will guess half the time right and half the time wrong. This means that the probability that Bob will receive a wrong bit is $\frac{1}{2\sqrt{2}} \approx 35.5\%$. These errors can easily be detected if Alice and Bob add one more step to their protocol.

(4) Alice and Bob publicly compare some of their bits. If there are no errors there is no eavesdropper and they keep the remaining bits. If there are errors, in the range of 35%, there is likely to be an eavesdropper. They then simply throw out all bits and try again.

However, Eve can do better. Her goal, besides learning as much as possible about the key, is also to introduce as few errors as possible. There are unavoidable errors in any communication scheme, partly due to the imperfections of the communication channel and partly due to the imperfect detection. Eve's goal is to remain below this unavoidable noise level in order to avoid being detected. So, suppose she has intercepted the particle but she now chooses the minimum error strategy to determine which state was sent. Using Eq. (28), her error rate will now be $\frac{1}{2}(1 - \frac{1}{\sqrt{2}}) \approx 14.6\%$ which is much less than the error rate that she introduces if she uses the unambiguous discrimination strategy. In addition, she still learns the key with a fidelity of about 85%. However, even this rather low error rate can still be detected if Alice and Bob modify the last step of their protocol.

(5) Alice and Bob publicly compare some of their bits. If there are no errors there is no eavesdropper and they keep the remaining bits. If there are errors, in the range of 14%, there is likely to be an eavesdropper. They then simply throw out all bits and try again.

This requirement is much more stringent than the one in Step (iv) of the original protocol. It is still possible to detect the presence of an eavesdropper but the requirements on the channel quality and detector efficiency are much more demanding than in the case when Eve uses the same strategy as Bob. So, here we had an example where one state discrimination strategy is optimal for the intended recipient and the other for the eavesdropper and to analyze the worst case scenario for Alice and Bob we have to consider all of their possibilities. There are many other QKD protocols but this one is perhaps the clearest example of how important optimal detection strategies are for quantum communication.

4. Recent developments

It is interesting to note that the two main discrimination strategies have evolved very differently from the beginning. Minimum error discrimination (also known as hypothesis testing or best guess discrimination) considered pure and mixed states on equal footing from its introduction. However, it is very difficult to generalize this strategy to more than two states, except for some very special highly symmetrical cases. Unambiguous discrimination, on the other hand, dealt with pure states for a long time after it was introduced. It is precisely the area of the unambiguous discrimination of mixed states that has seen a significant amount of progress recently.

Optimum unambiguous discrimination between two mixed states is an issue of ongoing theoretical research[15–24]. In contrast to minimum-error discrimination, there does not exist a compact formula expressing the minimum probability of inconclusive results, i. e. the minimum failure probability, for unambiguously discriminating two mixed states that are completely arbitrary. However, analytical solutions can be obtained for certain special classes of density operators, including the cases that are of interest for this review.

Unambiguous discrimination involving a mixed state started with Sun *et al.*[15] where the complete optimal solution for discriminating a pure state from a rank 2 mixed state was given. This later was generalized to the optimal unambiguous discrimination of a pure state and a mixed state of arbitrary rank[16].

In a seminal paper Rudolph, Spekkens and Turner[17] clearly established the principles behind the unambiguous discrimination of mixed states. In order to understand their reasoning, we need to introduce some terminology. The support of a density operator is defined as the subspace spanned by its eigenvectors belonging to nonzero eigenvalues The kernel of the density operator is the orthogonal complement to its support. Any measurement in the kernel of a state with a positive outcome unambiguously identifies the other state. Of course, this is not necessarily optimal. In the same paper it has also been shown that the fidelity of the two density operators represents an absolute lower bound for the failure probability of the unambiguous discrimination. They have also provided an analytical solution for the special case of two density matrices of rank $(d-1)$ that jointly span a d dimensional Hilbert space.

Reduction theorems were derived by Raynal *et al.*[18] that allow one to reduce the general problem of optimal unambiguous discrimination of two arbitrary density matrices of rank r_1 and r_2, respectively, to a standard form. This standard problem can be formulated as optimal unambiguous discrimination of two density matrices of the same rank d that together span a $2d$ dimensional Hilbert space. d is bounded by the smaller of the two ranks, $d \leq min(r_1, r_2)$. It is, therefore, sufficient to study only problems that are given in standard form.

Built on these developments, Herzog *et al.*[22] derived general inequalities for the fidelity bound and showed that there exist density operators for which it can never be saturated. In a closely related paper we have presented exact analytical solutions for the special case[24] when the two density operators in standard form are diagonal in the canonical basis, *i.e.* their spectral representation coincides with the Jordan basis and found explicit conditions for when the fidelity bound can be saturated. The solution found in these two papers also applies to the comparison of two pure states[25], also with arbitrary prior probabilities[22,26]. Moreover, it also applies to programmable discrimination[27] that distinguishes between two pure states when one[28] or both of them[27–30] are unknown and the discrimination is performed with reference copies.

In view of the reduction theorems one can say that in the group of solutions listed above the optimum measurement is either equivalent to that of discriminating a pair of pure states or can be constructed from the solutions for discriminating pairs of pure states. Recently, there has been progress in moving beyond cases that can be reduced to pure state discrimination. Raynal and Lütkenhaus[23] derived general expressions for optimum measurement operators which hold in the special case that for the given prior probabilities of the states the lower bound of failure probability is saturated. Herzog[31] and Raynal and Lütkenhaus[32] found the complete solution for the case of geometrically uniform (GU) mixed states, that is for the case when the two density matrices to be discriminated are related by a unitary transformation, $\rho_2 = U\rho_1 U^\dagger$, $U^2 = I$ where I is the identity in a four dimensional Hilbert space. As a consequence of the geometric uniformity, density matrices of this sort have identical spectra. Interestingly, by extending the B92 protocol[7] based on two non-orthogonal pure states to the case of two mixed states, Koashi and Imoto[33] found that for the communication to be secure the two density operators must belong to this class.

5. Summary and outlook

In the Appendices of this paper we have presented a tutorial review of the theory of generalized measurements (Positive Operator Valued Measures, POVMs). In our treatment the POVM emerges as a natural and for most practical purposes completely satisfactory generalization of the standard projector valued quantum measurement, originally introduced by von Neumann. Based on Neumark's Theorem, we have shown a general method to implement POVMs experimentally.

Based on the theory outlined in the Appendices, we have shown that optimal state discrimination problems are equivalent to finding measurement strategies that optimize some reasonably chosen figure of merit. The solution often leads to POVMs and we gave a thorough overview of the two most commonly employed state discrimination strategies, discrimination with minimum error (ME) and optimal unambiguous state discrimination (UD).

As an application we have analyzed the performance of the B92 quantum key distribution protocol and have shown that the optimal measurement strategies for Bob, the intended recipient of the key and Eve, the eavesdropper, are different. In fact it is the UD strategy for Bob and the ME strategy for Eve that are found optimal, at least against individual attacks.

Most of the analytical results pertaining to the unambiguous discrimination of two quantum states consider pure states. We have concluded our paper with a brief overview of recent progress in the area of optimum unambiguous discrimination of two mixed states. There are solutions to special cases but the general case is still an open problem.

In summary, state discrimination is a rapidly developing subfield of quantum information theory, touching the very foundation of quantum mechanics. Perhaps the most immediate open problem is the unambiguous discrimination of two mixed states. Although the solution is known for many special cases, no closed form solution is available for two general mixed states. It is fully expected that the optimal unambiguous discrimination of two rank 2 density matrices will be solved in the near future. However, no analytical expression is expected for higher rank problems due to the fact that the dependence on the relevant parameters is highly nonlinear. On the other hand, the general case is an example of a semidefinite programming problem[19] and as such, very efficient numerical methods can be employed to find a numerically optimized solution to the underlying measurement problem.

Acknowledgements

This research was partially supported by a grant from the Humboldt Foundation and from PSC-CUNY. The author is grateful to Ulrike Herzog, E. Feldman, and M. Hillery for many helpful discussions and fruitful collaborations.

Appendix A. Standard quantum measurements

Measurements are an integral part of quantum information processing. Reading out the quantum information at the end of the processing pipeline is equivalent to learning what final state the system is in at the output since information is encoded in the state. Since finding out the state of a system can be done only by performing measurements on it, we need a thorough understanding of the quantum theory (and practice) of measurements. To this end we will begin by a brief review of the postulates of standard quantum measurement theory, due essentially to von Neumann. Then, by analyzing the underlying assumptions we will show that some of the postulates can be replaced by more relaxed ones and this will lead us to the concept of generalized measurements (Positive Operator Valued Measures, POVMs) which are particularly useful in measurement optimization problems. Next, by invoking Neumark's theorem we will show how to actually implement POVMs experimentally. These general concepts will be illustrated in the next section by explicitly working out optimized measurement schemes for various state discrimination strategies.

We begin here with a brief summary of the postulates of standard quantum measurements. Standard, or projective, measurements were introduced by von Neumann [34] by analyzing a simple model for the coupling between the system to be measured and the meter and by appropriately generalizing the predictions of the model. Deatails of this analysis are omitted here, for the sake of brevity.

Without losing too much of the generality, we assume that the Hilbert space is finite and discrete, in order to keep the notation simple. Then the postulates read as

0. To every observable in quantum mechanics there corresponds a Hermitian operator X which has the spectral representation $X = \sum_j \lambda_j |j\rangle\langle j|$. From the hermiticity of X it follows that the eigenvalues λ_j are real. For simplicity we assume that the eigenvalues are nondegenerate and the corresponding eigenvectors, $\{|j\rangle\}$, form a complete orthonormal basis set.

1. The projectors $P_j = |j\rangle\langle j|$ span the entire Hilbert space, $\sum_j P_j = 1$.

2. From the orthogonality of the states we have $P_i P_j = P_i \delta_{ij}$. In particular, $P_i^2 = P_i$ from where it follows that the eigenvalues of any projector are 0 and 1.

3. A measurement of X yields one of the eigenvalues λ_j.

4. The state of the system after the measurement is $|\phi_j\rangle = \frac{P_j|\psi\rangle}{\sqrt{\langle\psi|P_j|\psi\rangle}}$ if the outcome is λ_j.

5. The probability that this particular outcome is found as the measurement result is $p_j = ||P_j\psi\rangle||^2 = \langle\psi|P_j^2|\psi\rangle = \langle\psi|P_j|\psi\rangle$ where we used the property 2.

6. If we perform the measurement but we do not record the results, the postmeasurement state can be described by the density operator $\rho =$

$\sum_j p_j|\phi_j\rangle\langle\phi_j| = \sum_j P_j|\psi\rangle\langle\psi|P_j$.

These seven postulates adequately describe what happens to the system during the measurement if it was initially in a pure state. If the system is initially in the mixed state ρ the last three postulates are to be replaced by their immediate generalizations:

4a. The state of the system after the measurement is $\rho_j = \frac{P_j\rho P_j}{Tr(P_j\rho P_j)} = \frac{P_j\rho P_j}{Tr(P_j\rho)}$ if the outcome is λ_j.

5a. The probability that this particular outcome is found as the measurement result is $p_j = Tr(P_j\rho P_j) = Tr(P_j^2\rho) = Tr(P_j\rho)$ where, again, we used the property 2.

6a. If we perform the measurement but we do not record the results, the postmeasurement state can be described by the density operator $\tilde{\rho} = \sum_j p_j\rho_j = \sum_j P_j\rho P_j$.

Of course, $4a$–$6a$ reduce to 4–6 for the pure state density matrix $\rho = |\psi\rangle\langle\psi|$. Therefore, in what follows we use the density matrix to describe a general (pure or mixed) quantum state unless we want to emphasize that the state is pure.

Let us summarize the message of these postulates. They essentially tell us that the measurement process is random, we cannot predict its outcome. What we can predict is the spectrum of the possible outcomes and the probability that a particular outcome is found in an actual measurement. This leads us to the ensemble interpretation of quantum mechanics. The state $|\psi\rangle$ (or ρ for mixed states) describes not a single system but an ensemble of identically prepared systems. If we perform the same measurement on each member of the ensemble we can predict the possible measurement results and the probabilities with which they occur but we cannot predict the outcome of an individual measurement. Except, of course, when the probability of a certain outcome is 0 or 1. With the help of these postulates we can then calculate the moments of the probability distribution, $\{p_j\}$, generated by the measurement. The first moment is the average of a large number of identical measurements performed on the initial ensemble. It is called the expectation value of X and is denoted as $\langle X\rangle$,

$$\langle X\rangle = \sum_j \lambda_j p_j = \sum_j \lambda_j Tr(P_j\rho) = Tr(X\rho) , \tag{A.1}$$

where we used the spectral representation of X. The second moment,

$Tr(X^2\rho)$, is related to the variance,

$$\langle (X - \langle X \rangle)^2 \rangle = Tr(X^2\rho) - \langle X \rangle^2 \ . \tag{A.2}$$

Higher moments can also be calculated in a straightforward manner but typically the first and second moments are the most important ones to consider.

Appendix B. Positive Operator Valued Measures (POVMs)

Now we are in the position to put the postulates of standard measurement theory under closer scrutiny. What the last three postulates provide us with is, in fact, an algorithm to generate probabilities. The generated probabilities are non-negative, $0 \leq p_j \leq 1$, and the probability distribution is normalized to unity, $\sum_j p_j = 1$ which is a consequence of the first two postulates. Furthermore, the number of possible outcomes is bounded by the number of terms in the orthogonal decomposition of the identity operator of the Hilbert space. Obviously, one cannot have more orthogonal projections than the dimensionality, N_A, of the Hilbert space of the system, so $j \leq N_A$. It would, however, be often desirable to have more outcomes than the dimensionality while keeping the positivity and normalization of the probabilities. We will first show that this is formally possible: if we relax the above rather restrictive postulates and replace them with more flexible ones we can still obtain a meaningful probability generating algorithm. Then we will show that there are physical processes that fit these more general postulates.

Let us begin with the formal considerations and take a closer look at Postulate $5a$ (or 5) which is the one that gives us the prescription for the generation of probabilities. We notice that in order to get a positive probability by this prescription it is sufficient if P_j^2 is a positive operator, we do not need to require the positivity of an underlying P_j operator. So let us try the following. We introduce a positive operator, $\Pi_j \geq 0$, which is the generalization of P_j^2, and prescribe $p_j = Tr(\Pi_j \rho)$. Of course, we want to ensure that the probability distribution generated by this new prescription is still normalized. Inspecting the postulates we can easily figure out that normalization is a consequence of Postulate 1 and, therefore, require that $\sum_j \Pi_j = I$, that is the positive operators still represent a decomposition of the identity. We will call a decomposition of the identity in terms of positive operators, $\sum_j \Pi_j = I$, a POVM (Positive Operator Valued Measure) and $\Pi_j \geq 0$ the elements of the POVM. These generalizations will form the core of our new postulates $2'$ and $6'$.

As observed in the previous paragraph, for a POVM to exist we do not have to require orthogonality and positivity of the underlying P_j operators. Therefore, the underlying operators that, via Postulates 4 (or $4a$) and 6 (or $6a$), determine the postmeasurement state can be just about any operators, even non-hermitian ones. For projective measurements orthogonality was essentially a consequence of Postulate 2, which was our most constraining postulate because it restricted the number of terms in the decomposition of the identity to at most the dimensionality of the system. Let us now see how far we can get by abandoning it.

If we abandon Postulate 2 then the operators that generate the probability distribution are no longer the same as the ones that generate the postmeasurement states and we have a considerable amount of freedom in choosing them. Let us denote the operators that generate the postmeasurement state by A_j, they are the generalizations of the orthogonal projectors, P_j. In other words, we define the non-normalized postmeasurement state by $A_j|\psi\rangle$ and the corresponding normalized state after the measurement by $|\phi\rangle = A_j|\psi\rangle/\sqrt{\langle\psi|A_j^\dagger A_j|\psi\rangle}$. This expression will form the essence of our new Postulate 4′. It immediately tells us that Π_j has the structure $\Pi_j = A_j^\dagger A_j$ which by construction is a positive operator. Let us now use our freedom in designing the postmeasurement state. First note that, since the POVM elements are positive operators, $\Pi_j^{1/2}$ exists. Obviously, this is a possible choice for A_j. So is

$$A_j = U_j \Pi_j^{1/2} \ , \tag{B.1}$$

where U_j is an arbitrary unitary operator. This is the most general form of the detection operators, satisfying $A_j^\dagger A_j = \Pi_j$ and the above expression corresponds to their polar decomposition. What we see is that the POVM elements determine the absolute value operator through $|A_j| = \Pi_j^{1/2}$ but leave its unitary part open. The A_j operators represent a generalization of the projectors P_j whereas Π_j is a generalization of P_j^2. The set $\{A_j\}$ is called the set of detection operators and these operators figure prominently in our new postulates $2', 4'$ and $6'$ replacing the corresponding ones of the standard measurements.

With this we completed our goal that we set out to do at the beginning of this section, namely the generalization of all of the postulates of the standard measurement theory to more flexible ones while keeping the spirit of the old ones. What we see is that Postulate 0. is no longer necessary and rest of the new postulates read as

1′. We consider the decomposition of the identity, $\sum_j \Pi_j = 1$, in terms of positive operators, $\Pi_j \geq 0$. Such a decomposition is called a POVM (Positive Operator Valued Measure) and the Π_j the elements of the POVM.

2′. The elements of the POVM, Π_j, can be expressed in terms of the detection operators A_j as $\Pi_j = A_j^\dagger A_j$ where, in general, the detection operators are non-hermitian ones, restricted only by the requirement $\sum_j A_j^\dagger A_j = I$. Then, by construction, the POVM elements are positive operators.

3′. A detection yields one of the alternatives corresponding to an element of the POVM.

4′. The state of the system after the measurement is $|\phi\rangle = A_j|\psi\rangle/\sqrt{\langle\psi|A_j^\dagger A_j|\psi\rangle}$ if it was initially in the pure state $|\psi\rangle$, and $\rho_j = \frac{A_j \rho A_j^\dagger}{Tr(A_j \rho A_j^\dagger)} = \frac{A_j \rho A_j^\dagger}{Tr(A_j^\dagger A_j \rho)}$ if it was initially in the mixed state ρ.

5′. The probability that this particular alternative is found as the measurement result is $p_j = Tr(A_j \rho A_j^\dagger) = Tr(A_j^\dagger A_j \rho) = Tr(\Pi_j \rho)$ where we used the cyclic property of the trace operation.

6′. If we perform the measurement but we do not record the results, the postmeasurement state is described by the density operator $\tilde{\rho} = \sum_j p_j \rho_j = \sum_j A_j \rho A_j^\dagger$.

Very often we are not concerned with the state of the system after such operation is performed but only with the resulting probability distribution. For this, it is sufficient to consider Postulates 1′ and 5′ defining the probability of finding alternative j as the detection result. Note, that at no step did we require the orthogonality of the Π_j's. Since orthogonality is no longer a requirement, the number of terms in this decomposition of the identity is not bounded by N_A. In fact, the number of terms can be arbitrary. Obviously, what we arrived at is a generalization of the von Neumann projective measurement. It is a surprising generalization as it tells us that just about any operation that satisfies Postulates 1′ and 2′ is a legitimate operation that generates a valid probability distribution. It is a rather natural generalization of the standard quantum measurement, as it provides us with a well-defined algorithm that generates a well-behaved probability distribution. So this procedure can be regarded as a *generalized measurement* and, indeed, for most purposes it is a sufficient generalization of the standard quantum measurement.

A further note is in place here. Very often a projector projects on a one dimensional subspace, which is spanned by the vector $|\omega_j\rangle$, in which case it can be written as $P_j = |\omega_j\rangle\langle\omega_j|$. The corresponding generalization to a non-hermitian detection operator can be written as $A_j = c_j|\tilde{\omega}_j\rangle\langle\omega_j|$ where $\langle\omega_j|\omega_j\rangle = \langle\tilde{\omega}_j|\tilde{\omega}_j\rangle = 1$, and c_j is a complex number inside the unit circle $|c_j|^2 \leq 1$, and $\langle\omega_j|\tilde{\omega}_j\rangle$ is arbitrary. Then

$$\Pi_j = |c_j|^2|\omega_j\rangle\langle\omega_j| \ , \tag{B.2}$$

but $\langle\omega_j|\omega_k\rangle \not\sim \delta_{ij}$ and, hence, it is explicit that the POVM is not an orthogonal decomposition of the identity. Since $A_j|\psi\rangle = c_j\langle\omega_j|\psi\rangle|\tilde{\omega}_j\rangle$, we see that $|\tilde{\omega}_j\rangle$ is proportional to the postmeasurement state $|\phi_j\rangle$, and $p_j = \langle\psi|\Pi_j|\psi\rangle = \langle\psi|A_j^\dagger A_j|\psi\rangle = |c_j|^2|\langle\omega_j|\psi\rangle|^2$, we have that

$$|c_j|^2 = \frac{p_j}{|\langle\omega_j|\psi\rangle|^2} \ , \tag{B.3}$$

and

$$\Pi_j = \frac{p_j}{|\langle\omega_j|\psi\rangle|^2}|\omega_j\rangle\langle\omega_j| \ . \tag{B.4}$$

Of course, up to this point all this is just a formal mathematical generalization of the standard quantum measurement. The important question is: How can we implement such a thing physically? In the next section we set out to answer this question and then we will study examples of POVMs.

Appendix C. Neumark's theorem and the implementation of a POVM via generalized measurements

First, let us take a look at what happens if we couple our system to another system called ancilla, let them evolve, and then measure the ancilla. The Hilbert space of the system is $\mathcal{H}_A$ and the Hilbert space of the ancilla is $\mathcal{H}_B$. We want to gain information about the state of the system that we now denote as $|\psi_A\rangle$. Let $\{|m_B\rangle\}$ be an orthonormal basis for $\mathcal{H}_B$, and U_{AB} a unitary operator acting on $\mathcal{H}_A \otimes \mathcal{H}_B$. The probability p_m of measuring $|m_B\rangle$ is then given by

$$p_m = \| \ (I_A \otimes |m_B\rangle\langle m_B|)U_{AB}(|\psi_A\rangle \otimes |\phi_B\rangle) \ \|^2 \ . \tag{C.1}$$

Define

$$A_m|\psi_A\rangle \equiv \langle m_B|U_{AB}(|\psi_A\rangle \otimes |\phi_B\rangle) \ . \tag{C.2}$$

Then A_m is a linear operator on $\mathcal{H}_A$ that depends on $|m_B\rangle$, $|\phi_B\rangle$ and U_{AB}. With the help of this definition we can write the measurement probability as

$$p_m = \| A_m|\psi_A\rangle \otimes |m_B\rangle \|^2 = \langle\psi_A|A_m^\dagger A_m|\psi_A\rangle \ . \tag{C.3}$$

Note that

$$\begin{aligned}\sum_m \langle\psi_A|A_m^\dagger A_m|\psi_A\rangle &= \sum_m (\langle\psi_A| \otimes \langle\phi_B|)U_{AB}^\dagger|m_B\rangle\langle m_B|U_{AB}(|\psi_A\rangle \otimes |\phi_B\rangle) \\ &= 1 \ . \end{aligned}\tag{C.4}$$

Since this is true for any $|\psi_A\rangle$, we must have that

$$\sum_m A_m^\dagger A_m = I_A \ , \tag{C.5}$$

where I_A is the identity in $\mathcal{H}_A$.

The non-normalized state of the total 'system plus ancilla' after the measurement is $A_m|\psi_A\rangle \otimes |m_B\rangle$ so the (normalized) postmeasurement state of the system alone is

$$|\phi_A\rangle = \frac{1}{\sqrt{\langle\psi_A|A_m^\dagger A_m|\psi_A\rangle}} A_m|\psi_A\rangle \ . \tag{C.6}$$

The set $\{A_m^\dagger A_m\}$ thus gives a decomposition of the identity in terms of positive operators. Therefore, we can identify them with a POVM. In fact, what we see here is the first half of Neumark's theorem: If we couple our system to an ancilla, let them evolve so that they become entangled, and perform a measurement on the ancilla, which collapses the ancilla to one of the basis vectors of the ancilla space, then this procedure will also transform the system because the ancilla degrees of freedom are now entangled to the system. The transformation of the state of the system is, however, neither unitary nor a projection. It can adequately be described as a POVM so the above procedure corresponds to a POVM in the system Hilbert space. Thus, we have just found a procedure that, when we look at the system only, looks like a POVM. We now know that there are physical processes that can adequately be described as POVMs.

Next we address the question, given the set of operators $\{A_m\}$ acting on $\mathcal{H}_A$ such that $\sum_m A_m^\dagger A_m = I$, can this be interpreted as resulting from a measurement on a larger space? That is, can we find $\mathcal{H} = \mathcal{H}_A \otimes \mathcal{H}_B$, $|\phi_B\rangle$, $\{|m_B\rangle\} \in \mathcal{H}_B$ and U_{AB} acting on $\mathcal{H}$ such that

$$A_m|\psi_A\rangle = \langle m_B|U_{AB}(|\psi_A\rangle \otimes |\phi_B\rangle) \tag{C.7}$$

holds?

The answer to this question is yes as we will now prove it constructively. Let us choose $\mathcal{H}_B$ to have dimension M and let $\{|m_B\rangle\}$ be an orthonormal basis for $\mathcal{H}_B$, and choose $|\phi_B\rangle$ to be an arbitrary state of $\mathcal{H}_B$. Let us further define a transformation U_{AB} via

$$U_{AB}(|\psi_A\rangle \otimes |\phi_B\rangle) = \sum_m A_m|\psi_A\rangle \otimes |m_B\rangle \ , \tag{C.8}$$

which implies the Eq. (C.7). U_{AB} is inner product preserving,

$$(\sum_{m'} \langle\psi'_A|\otimes\langle m'_B A^{\dagger}_{m'})(\sum_m A_m|\psi_A\rangle\otimes|m_B\rangle) = \sum_m \langle\psi'_A|A^{\dagger}_m A_m|\psi_A\rangle = \langle\psi'_A|\psi_A\rangle \ , \tag{C.9}$$

so it is unitary on the one-dimensional subspace spanned by $|\phi_B\rangle$ and it can be extended to a full unitary operator on $\mathcal{H}_A \otimes \mathcal{H}_B$ because, e.g., on the subspace that is orthogonal to $|\phi_B\rangle$ it can be the identity.

This completes the proof of Neumark's theorem which asserts that there is a one-to-one correspondence between a POVM and the above procedure which sometimes is itself called a generalized measurement. Hence, a generalized measurement can be regarded as the physical implementation of a given POVM. As a final remark to this Section we note that there are two standard methods to extend a Hilbert space. One is the tensor product extension, the other is the direct sum extension. Both methods are employed in practical schemes and they lead to rather different implementations of the same POVM. We will not pursue this issue further here but details can be found in Ref. [4], for example.

References

1. M. A. Nielsen and I. L. Chuang, *Quantum Computation and Information* (Cambridge University Press, 2000).
2. A. Peres, *Quantum Theory: Concepts and Methods* (Kluwer, 1995).
3. A. Chefles, "Quantum state discrimination", Contemporary Physics **41**, 401 (2000).
4. J. A. Bergou, U. Herzog, and M. Hillery, "Quantum state discrimination", in *Quantum State Estimation* (Lect. Notes Phys. vol 649), M. Paris and J. Řehaček, eds. (Springer, 2004), p. 417.
5. C. W. Helstrom, *Quantum Detection and Estimation Theory* (Academic Press, 1976).
6. A. S. Holevo, *Probabilistic and statistical aspects of quantum theory* (North-Holland Publishing Company, 1979).
7. C. H. Bennett, "Quantum cryptography using any two nonorthogonal states", Phys. Rev. Lett. **68**, 3121 (1992).

8. I. D. Ivanovic, "How to differentiate between nonorthogonal states", Phys. Lett. A **123**, 257 (1987).
9. D. Dieks, "Overlap and distinguishability of quantum states", Phys. Lett. A **126**, 303 (1988).
10. A. Peres, "How to differentiate between two nonorthogonal states", Phys. Lett. A **128**, 19 (1988).
11. G. Jaeger and A. Shimony, "Optimal distinction between two non-orthogonal quantum states", Phys. Lett. A **197**, 83 (1995).
12. Y. Sun, M. Hillery, and J. A. Bergou, "Optimum unambiguous discrimination between linearly independent nonorthogonal quantum states and its optical realization", Phys. Rev. A **64**, 022311 (2001).
13. M. Mohseni, A. M. Steinberg, and J. A. Bergou, " Optical Realization of Optimal Unambiguous Discrimination for Pure and Mixed Quantum States", Phys. Rev. Lett. **93**, 200403 (2004).
14. U. Herzog and J. A. Bergou, "Distinguishing mixed quantum states: Minimum-error discrimination versus optimum unambiguous discrimination", Phys. Rev. A **70**, 022302 (2004).
15. Y. Sun, J. A. Bergou, and M. Hillery, Phys. Rev. A **66**, 032315 (2002).
16. J. A Bergou, U. Herzog, and M. Hillery, Phys. Rev. Lett **90**, 257901. (2003)
17. T. Rudolph, R. W. Spekkens, and P. S. Turner, "Unambiguous discrimination of mixed states", Phys. Rev. A **68**, 010301(R) (2003).
18. Ph. Raynal, N. Lütkenhaus, and S. J. van Enk, "Reduction theorems for optimal unambiguous state discrimination of density matrices", Phys. Rev. A **68**, 022308 (2003).
19. Y. Eldar, "A Semidefinite Programming Approach to Optimal Unambiguous Discrimination of Quantum States", IEEE Trans. Inf. Theory **49**, 446 (2003).
20. Y. Feng, R. Duan, and M. Ying, "Unambiguous discrimination between mixed quantum states", Phys. Rev. A **70**, 012308 (2004).
21. Y. C. Eldar, M. Stojnic, and B. Hassibi, "Optimal quantum detectors for unambiguous detection of mixed states", Phys. Rev. A **69**, 062318 (2004).
22. U. Herzog and J. A. Bergou, "Optimum unambiguous discrimination of two mixed quantum states", Phys. Rev. A **71**, 050301(R) (2005).
23. Ph. Raynal and N. Lütkenhaus, "Optimal unambiguous state discrimination of two density matrices: Lower bound and a class of exact solutions", Phys. Rev. A **72**, 022342 (2005).
24. J. A. Bergou, E. Feldman, and M. Hillery, "Optimal unambiguous discrimination of two subspaces as a case in mixed-state discrimination", Phys. Rev. A **73**, 032107 (2006).
25. S. M. Barnett, A. Chefles, and I. Jex, Phys. Let. A **307**, 189 (2003).
26. M. Kleinmann, H. Kampermann, and D. Bruß, Phys. Rev. A **72**, 032308 (2005).
27. J. A. Bergou and M. Hillery, "Universal programmable quantum state discriminator that is optimal for unambiguously distinguishing between unknown states", Phys. Rev. Lett. **94**, 160501 (2005).
28. J. A. Bergou, V. Bužek, E. Feldman, U. Herzog, and M. Hillery, "Programmable quantum-state discriminators with simple programs", Phys. Rev.

A **73**, 062334 (2006).
29. A. Hayashi, M. Horibe, and T. Hashimoto, "Quantum pure-state identification", Phys. Rev. A **72**, 052306 (2005).
30. A. Hayashi, M. Horibe, and T. Hashimoto, "Unambiguous pure-state identification without classical knowledge", Phys. Rev. A **73**, 012328 (2006).
31. U. Herzog, "Optimum unambiguous discrimination of two mixed states and application to a class of similar states", Phys. Rev. A **75**, 052309 (2007).
32. Ph. Raynal and N. Lütkenhaus, "Optimal unambiguous state discrimination of two density matrices: A second class of exact solutions", Phys. Rev. A **76**, 052322 (2007).
33. M. Koashi and M. Imoto, "Quantum cryptography based on two mixed states", Phys. Rev. Lett. **77**, 2137 (1996).
34. J. von Neumann, *The Mathematical Foundations of Quantum Mechanics* (Princeton University Press, 1995)

Physical problems of brain-computer interfacing

Peter Fromherz
Department of Membrane and Neurophysics
Max Planck Institute for Biochemistry
Martinsried / Munich, Germany

The direct electrical interfacing of semiconductor chips with individual nerve cells and with brain tissue is considered. At first, the structure of the cell-chip contact is described and then the electrical coupling is characterized between ion channels, the electrical elements of nerve cells, and transistors and capacitors of silicon chips. On that basis the signal transmission between microelectronics and the microionics of nerve cells is implemented in both directions. Simple hybrid neuroelectronic systems are assembled with neuron pairs and small neuronal networks. Finally, the interfacing is extended to brain tissue. The application of CMOS chips with recording and stimulation sites allows an imaging of neuronal activity with high spatiotemporal resolution. Goal of the work in an integration of neuronal network dynamics and digital electronics on a microscopic level for applications in brain research, medical prosthetics and information technology

1. Introduction

Computers and brains work electrically. However, their charge carriers are different, electrons in solid silicon and ions in liquid water. It is an intellectual and technological challenge to join these different systems right on the level of their electronic and ionic signals. In principle, Luigi Galvani has established an electrical coupling between inorganic solids and excitable living tissue in the 18th century. Today, after fifty years of dramatic developments in semiconductor technology as well as in neurobiology and molecular biology, we may envisage a most complex integration of microionics and microelectronics with myriads of nerve cells and semiconductor devices and breathtaking applications in medicine and information technology.

Usually, in neurophysiology the coupling of electronic and ionic signals is achieved with perfectly unpolarized solid/water contacts such as Ag/AgCl electrodes where ionic and electronic currents are transformed into each other.That approach is not suitable for an iono-electronic interfacing on a microscopic scale with numerous contacts. Interfacing must be implemented with perfectly polarized solid/water contacts without Faradayic current. Semiconductor chips must be protected from corrosion, nerve cells from electrochemical reaction products. Thus the communication between microionics and microelectronics must be achieved by displacement currents across an insulating interface [1].

The present paper describes step by step the mechanisms of neuroelectronic interfacing on the scale of nanometers, micrometers and millimeters. It starts with the contact between an individual nerve cell and a silicon chip. Then the interfacing of the fundamental devices of the brain and the fundamental devices of computers is studied. On that basis, the interfacing of nerve cells with capacitors and transistors and the assembly of neuroelectronic hybrids with two nerve cells and with simple neuronal networks is considered. Finally, the challenges with electronic interfacing of brain tissue are addressed.

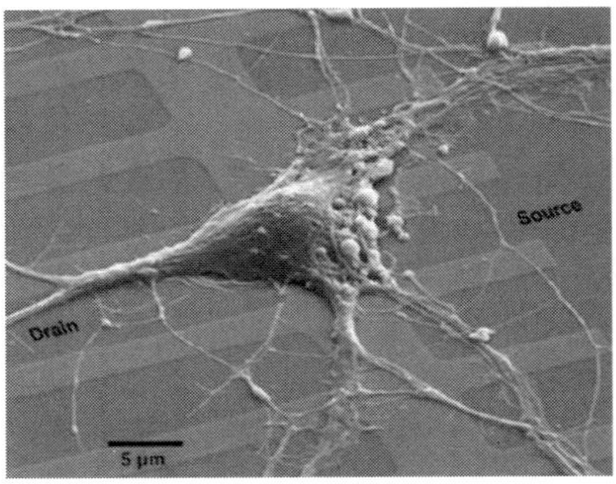

Figure 1. Nerve cell from rat brain on silicon chip. Electronmicrograph after fixation. The surface of the chip consists of silicon dioxide. The metalfree gates of a linear array of field-effect transistors are visible as bright squares between the dark regions of sources and drains. The neuron is cultured for several days in electrolyte.

2. Cell-Chip Contact

A simple hybrid with a nerve cell from rat brain and a transistor is depicted in Fig. 1. The cell is surrounded by a membrane with an electrically insulating core of lipid. That lipid bilayer (thickness about 5 nm) separates the bath with a concentration of 150 mM sodium ions from the cytoplasm with about 150 mM potassium ions. The silicon is coated with silicon dioxide (thickness 10 nm). The basic problem with respect to the electrical interaction of cell and chip is the structure of the contact between the lipid bilayer and the oxide: What is their distance? (ii) What is the electrical resistance inbetween ?

2.1. *Distance*

Silicon reflects light such that standing modes of the electromagnetic field are formed in front of its surface. We take advantage of that effect to measure the distance of cells and silicon using fluorescent dye molecules as antennas [2]. For that purpose, silicon chips are fabricated with microscopic oxide terraces

(2.5 μm x 2.5 μm, step height about 20 nm). Cells are cultured on these substrates. The terraces of defined height together with the unknown distance between chip and cell bring the cell membrane into different positions in the standing light modes. The lipid bilayer of the cell membrane is labeled with a fluorescent dye. When we evaluate the fluorescence intensities on a set of terraces with the classical antenna theory, we find that there is a cleft between the oxide and the lipid bilayer. Results for rat neurons on chips coated with polylysine are plotted in Fig. 2. The average distance is around 50 nm.

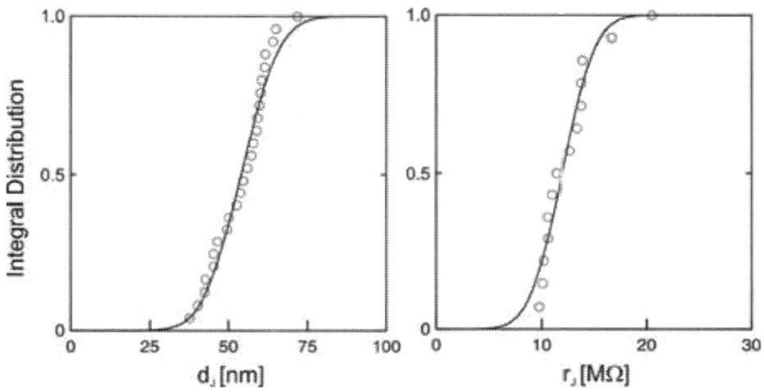

Figure 2. Contact between nerve cell and silicon chip coated with polylysine. Left: Integral distribution function of the distance between the lipid bilayer of the cell membrane and the oxide of the chip. Right: Integral distribution function of the sheet resistance between cell and chip.

2.2. *Resistance*

The electrical coupling of chip and cell depends on the resistance of the 50 nm cleft between oxide and membrane. To measure the sheet resistance we apply an alternating voltage to the chip. The resistance of the cleft together with the capacitances of oxide and membrane determines the phase of the alternating voltage across the membrane. We probe the phase shift with a fluorescent dye that exhibits a molecular Stark effect [3]. The phase map of the fluorescence intensity is evaluated in terms of a core-coat conductor theory in two dimensions. Results for rat neurons on polylysine are shown in Fig. 2. The average sheet resistance is around 10 MΩ/square.

When we combine the resistance measurement with the distance measurement, we find that the 50 nm cleft is filled with bulk electrolyte from the surrounding bath. That thin film of electrolyte between oxide and membrane prevents a direct electrostatic polarization of a cell from a chip and of a chip from a cell, respectively.

3. Ion-Electron Coupling

The electrical elements of nerve cells are voltage-gated ion channels. These molecules are embedded in the lipid bilayer of the membrane. They can be in an open and in a closed state. When they are open, they selectively transmit ionic current through the membrane, Na^+ inward current or K^+ outward current. The opening and closing of the channels is connected with a displacement of electrical charge across the membrane. As a consequence, opening and closing are controlled by the voltage across the membrane.

The electrical communication between semiconductors and cells must rely on an interaction of the ion channels with the chip. Two devices are used, electrolyte/oxide/silicon (EOS) capacitors for the activation of the channels and electrolyte/oxide/silicon (EOS) field-effect transistors for the detection of open channels. The interaction of channels and chips is determined by the electrical nature of the cell-chip contact with the insulating oxide, the insulating lipid bilayer and the thin film of electrolyte inbetween (Fig. 3).

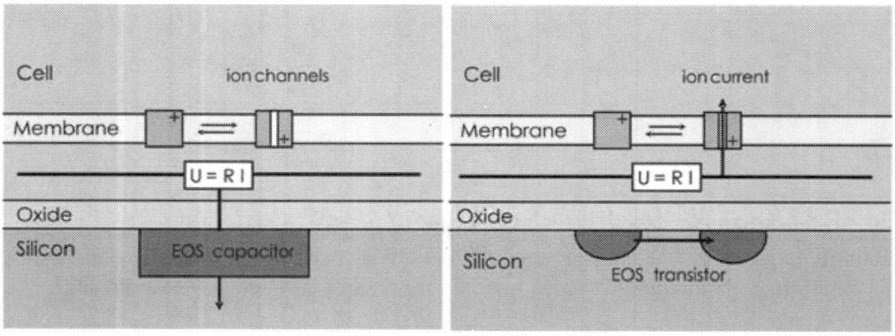

Figure 3. Ionoelectronic interfacing on the basis of Ohm's law. Left: Activation of voltage-gated ion channels by an electrolyte/oxide/semiconductor (EOS) capacitor. When a falling voltage ramp is applied to the chip, the displacement current across the oxide gives rise to a negative extracellular voltage that opens the channels. Right: Recording of ion channels by an EOS field-effect transistor. Inward ion current through open channels gives rise to a negative extracellular voltage such that the electronic current in a field-effect transistor is modulated.

3.1. *Ion Channels and EOS Capacitors*

We test the signaling from chip to cell with HEK293 cells that are transfected with the gene of the potassium channel Kv1.3. These cells are cultured on an EOS capacitor. A falling voltage ramp is applied to the capacitor (Fig. 3). A displacement current flows across the oxide and gives rise to a current along the sheet resistance of the cell-chip contact. The resulting negative extracellular voltage opens the channels in the attached membrane.

To achieve a sufficiently high displacement current, the chip is insulated by titanium dioxide with a high dielectric constant [4]. The ion current through the membrane is measured with a micropipette that is fused to the cell membrane. The experiment is performed at a constant intracellular voltage where the channels are closed without capacitive stimulation. During falling voltage ramps, the current through the membrane is enhanced, depending on the slope of the ramp [5]. The current signal is specific for the Kv1.3 channels as proven by the application of a selective toxin.

The experiment shows that the ionic current in the cell is controlled by an electronic signal in the semiconductor. Note the current-voltage-current mechanism: A displacement current across the oxide is transformed to an extracellular voltage between oxide and membrane that induces an ion current through the membrane.

3.2. *Ion Channels and EOS transistors*

We test the signaling from cells to chips with HEK293 cells that are transfected with the gene of a sodium channel. The cells are cultured on an EOS field-effect transistor (EOSFET). The thin electrolyte film between cell and chip plays the role of a gate (Fig. 3). Using a micropipette, positive voltage pulses are applied to the cell to open the channels. Na^+ current flows into the cell and along the sheet resistance of the cell-chip contact. It gives rise to a negative extracellular voltage that modulates the electronic current. We simultaneously record the transistor signal as well as the total membrane current using the micropipette. When positive voltage pulses are applied to the cell, a typical transient inward current is observed through the membrane. Simultaneously, a transient transistor signal appears that has a similar waveform. A quantitative evaluation shows that the transistor voltage is proportional to the membrane current as determined by the resistance of the cell-chip contact [6].

The experiment demonstrates that the electronic current in the semiconductor is controlled by the ionic signal in the cell. Note again the current-voltage-current mechanism: An ion current through the membrane is transformed into an extracellular voltage between oxide and membrane that modulates the electron current in the semiconductor.

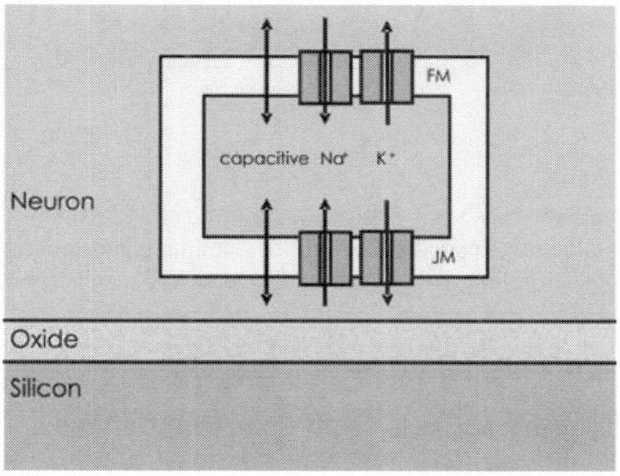

Figure 4. Nerve cell on a silicon chip. The current through various ion channels and the capacitive current must be considered through the attached membrane (JM) and the free upper membrane (FM) for stimulation with an EOS capacitor as well as for recording with an EOS transistor.

4. Nerve Cell on Semiconductor

A first step towards an integration of neuronal dynamics and integrated electronics is the interfacing of individual nerve cells and silicon microstructures. We consider the excitation of nerve cells by EOS capacitors and the recording of neuronal activity by EOS transistors. A crucial issue is the contribution of two domains of a nerve cell, the membrane that is attached to the chip and the upper free membrane in contact to the bath as illustrated in Fig. 4.

When a voltage ramp is applied to a capacitor, the extracellular voltage in the cleft between chip and cell polarizes the attached membrane as well as the upper membrane in opposite directions. Ion channels may be affected in both domains. On the other hand, when a nerve cell is excited, there are ion currents through the attached as well as through the upper free membrane that determine the change of intracellular voltage during an action potential. The current through the attached membrane gives rise to an extracellular voltage that modulates the transistor current. It has an ionic component due to the channels in the attached membrane and a capacitive component that depends on the total ion current through the attached and free membrane.

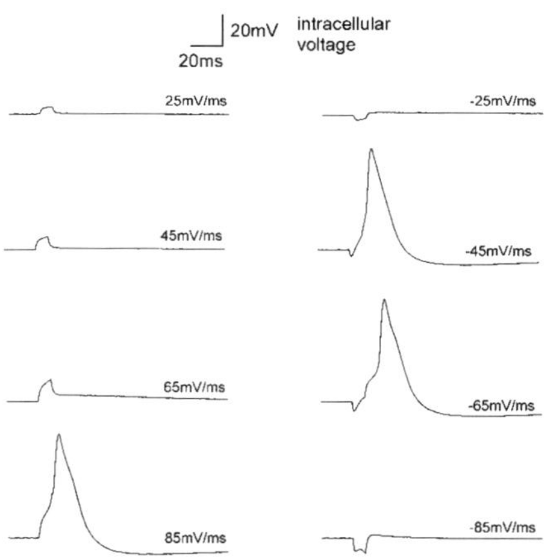

Figure 5. Capacitor stimulation of snail neuron. Left: intracellular voltage for different slopes of rising voltage ramps applied to a capacitor. The upper membrane is depolarized. Right: intracellular voltage for different slopes of falling voltage ramps. The attached membrane is depolarized.

4.1. *EOS Capacitor Stimulation*

We dissociate large nerve cells from snails and attach them to a capacitor. Voltage ramps are applied and the intracellular voltage is recorded with a micropipette [7]. A rising ramp leads to a positive extracellular voltage between chip and cell. The concomitant voltage drop across the upper membrane gives rise to an activation of Na^+ channels such that an action potential is elicited (Fig. 5). A rather high slope of the ramp is required because most of the extracellular voltage drops across the attached membrane with its small area. When a falling voltage ramp is applied, a negative extracellular voltage is induced between chip and cell. Na^+ channels are activated in the attached membrane. The resulting inward Na^+current depolarizes the whole cell membrane such that an action potential is elicited (Fig. 5). The effect disappears for strong stimuli when an outward K^+ current is activated.

4.2. *EOS FET Recording*

Nerve cells from the leech are placed on EOS field-effect transistors. They are impaled with a micropipette electrode for stimulation and for recording the intracellular voltage. The calibrated transistors detect the transient extracellular voltage between cell and chip. Three examples are shown in Fig. 6 [8]. In the first record, the extracellular voltage resembles the first derivative of the intracellular voltage. In the second record, the transistor signal shows a peak in the rising phase of the intracellular voltage and a trough in its falling phase.

The different signals reflect different contributions of ionic and capacitive currents through the attached cell membrane. The first signal is dominated by the capacitive current that is mainly driven by ionic currents of the upper membrane, whereas the second record is determined by ionic currents in the attached membrane with a sharp Na^+ inward current and a delayed K^+ outward current.

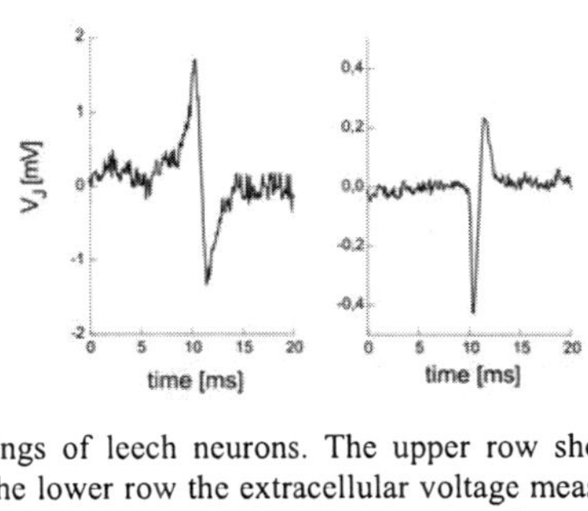

Figure 6. Transistor recordings of leech neurons. The upper row shows the intracellular voltage during an action potential, the lower row the extracellular voltage measured with an EOSFET. Left: the transistor voltage is dominated by capacitive current across the attached membrane,. Right: the signal is dominated by sodium and potassium currents across the attached membrane.

The different types of neuronal recordings indicate that the ion channels are inhomogeneously distributed in the cell membrane and that the cells are differently attached to the transistor in the two experiments.

That interpretation is tested when the neuronal excitation is not probed with a single EOS transistor, but with a closely packed array of sensor sites that are derived from the EOSFET principle. Such an array is fabricated by using an extended CMOS technology. Fig. 7 shows cultured snail neurons on the surface of a CMOS chip with 16384 sensor sites on one squaremillimeter. When a selected nerve cell is excited with a micropipette, a set of sensor sites beneath the cell is able to map the extracellular voltage in the cell-chip contact [9]. Different waveforms of the extracellular voltage are observed for different positions beneath a single nerve cell: Neuroelectronic interfacing depends on structural details of the cell-chip contact.

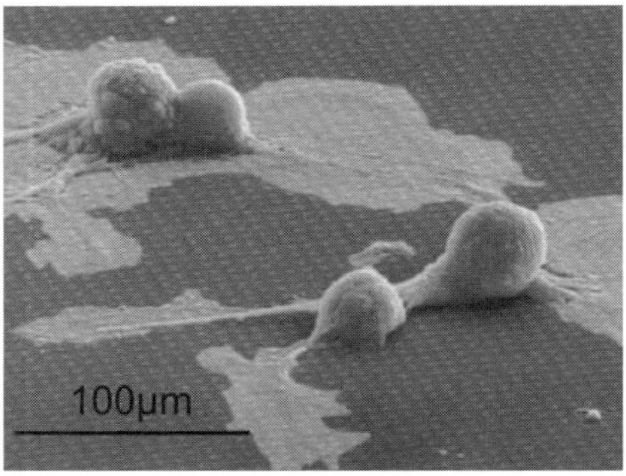

Figure 7. Nerve cells from snail on sensor transistors of a CMOS chip. Electronmicrograph. The contacts of recording sites (16384 on 1 squaremillimeter) shine through the TiO_2 surface of the chip. The sensor array allows a mapping of the extracellular voltage beneath individual nerve cells with a resolution of 7.8 micrometer.

4.3. *EOSFET Recording and Noise*

With small nerve cells from mammalian brains, we expect far smaller extracellular voltages during an action potential than for the large neurons from leech and snail. Successful recordings are achieved with low-noise EOS transistors that are fabricated in a buried-channel configuration. It is important to note that the background noise of transistors that are covered by the cell is larger than the noise of the free transistors. That effect is due to the Nyquist noise of the cleft resistance between cell and chip. In the frequency range of the action potentials, the power density of that thermal noise is similar to the intrinsic 1/f noise of the transistor. It sets a thermodynamic limit for the signal-to-noise ratio of neuroelectronic interfacing [10].

5. Elementary Neuroelectronic Hybrids

In a first step towards the assembly of neuroelectronic hybrids, we couple pairs of nerve cells to a chip with two different pathways as illustrated in Fig. 8.

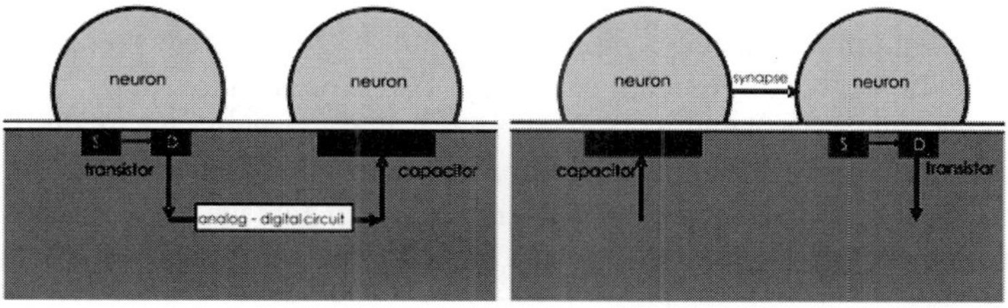

Figure 8. Neuroelectronic hybrids. Left: Cellular neuroprosthesis. The input neuron is probed by a transistor. The signal is processed on the chip and applied to a capacitor that stimulates the output neuron. Right: Neuronal memory on chip. The input neuron is stimulated from the chip and coupled by a synapse to the output neuron. The synaptic strength is enhanced by capacitive stimulation.

5.1. *Cellular Neuroprosthesis*

A kind of cellular neuroprosthesis is implemented in a neuron-silicon-silicon-neuron system where the activity of a neuron is recorded and used for stimulation of a second neuron after processing on the chip [11]. Two snail neurons are attached to a silicon chip with EOS capacitors and EOS transistors They have no direct contact. The spontaneous activity of the first neuron is recorded by a sensor transistor. The transistor record of an action potential is identified on the chip and elicits a digital signal. A delay line is triggered, and a burst of voltage pulses is applied to a stimulation capacitor such that neuronal activity is induced in the second neuron. A crucial problem in that system is the crosstalk from the stimulation EOS capacitor back to the EOS transistor. Resulting artifacts of the transistor record must be eliminated on the chip.

5.2. *Neuronal Memory on Chip*

A kind of neuronal memory element is obtained in a silicon-neuron-neuron-silicon system where a capacitively stimulated neuron is coupled to a second neuron through a chemical synapse and where the postsynaptic response is detected with a transistor [12]. Two identified snail neurons are attached to an array of EOS capacitors and EOS transistor such that they form a soma-soma contact with a chemical synapse. The presynaptic nerve cell is stimulated from a capacitor, the signal activates the chemical synapse and the postsynaptic excitation is recorded with an EOS transistor. In a first test, there may be no postsynaptic activity if the synaptic coupling is weak. When a series of capacitive presynaptic stimuli is applied, the synaptic strength is enhanced such that a second test reveals a postsynaptic action potential that is recorded by the transistor. The two-neuron hybrid implements a neuronal memory on chip.

6. Neuronal Networks on Chip

Complex neuronal networks rely (i) on a mapping between sets of neurons and (ii) on an enhanced synaptic strength by correlated presynaptic and postsynaptic activity (Hebbian learning). An experimental study of network dynamics requires (i) neuronal maps with a defined topology of the synaptic connections, and (ii) a noninvasive supervision of all neurons to induce learning and to observe the performance of the network. To achieve these goals we must control outgrowth and synapse formation, and we must fabricate silicon chips with arrays of many two-way contacts.

6.1. *Immobilized Neurons*

Neuronal networks can be obtained in cell culture from nerve cells that are dissociated from the brain of snails and rats. During outgrowth, however, the cell bodies are displaced on the chip such that the junctions with sensor transistors and stimulation capacitors are disrupted. That displacement is caused by the forces of the connecting neurites that tend to shorten in cell culture. To overcome that problem, picket fences of organic polymers are fabricated on chip and the neuronal cell bodies are mounted as shown in Fig. 9 [13]. The cell bodies are perfectly kept on the two-way contacts made of stimulation capacitors and sensor transistors, even after culturing for several days. In such a system, a selected neuron is excited by a capacitor, a postsynaptic action potential is elicited through electrical synapses and the signal is recorded with a transistor.

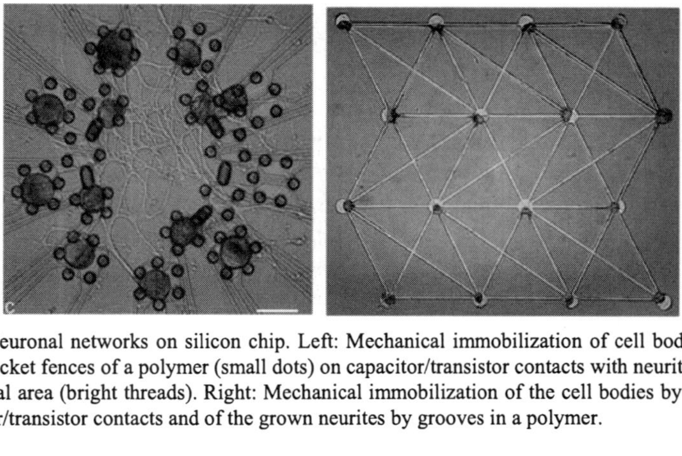

Figure 9. Neuronal networks on silicon chip. Left: Mechanical immobilization of cell bodies (dark blobs) by picket fences of a polymer (small dots) on capacitor/transistor contacts with neurites grown in the central area (bright threads). Right: Mechanical immobilization of the cell bodies by wells on 16 capacitor/transistor contacts and of the grown neurites by grooves in a polymer.

6.2. *Topographical Guidance*

It is well known that the outgrowth of neurons can be controlled by chemical patterns. However, that kind of guidance is not stable due to the strong contracting forces of growing neurites. Topographical guidance may overcome that problem. An example is shown in Fig. 9. Using an organic polymer, wells and connecting grooves are fabricated on a chip with 16 two-way contacts of stimulation capacitors and sensor transistors. After placing snail neurons into the wells, their outgrowth is guided by the grooves such that electrical synapses are formed [14]. In principle, it is possible to create defined neuronal networks that are supervised from a chip. However, at present the overall yield is too low to implement complex systems.

6.3. *Random Nets on CMOS Chips*

The problem of displaced neurons is overcome when an array of closely packed EOS capacitors and recording sites is used, such that a displaced neuron finds itself on one or more two-contact sites, wherever it is on the chip. Of course, the defined control of network geometry is sacrificed in such systems. However, when the changing geometry is continuously monitored by the chip, we may perform experiments with complex patterns of stimulation that induce a spatially distributed neuronal memory on chip.

6.4. *Synaptic Recording*

In order to supervise a neuronal network, it would be important not only to record the activity of all presynaptic and postsynaptic neurons, but also the activity of all synapses. In a first step towards that ambitious goal, we use chromaffin cells from the adrenal gland as a model for the presnaptic release of vesicles [15]. The cells are attached to EOSFETs. Upon chemical stimulation, a burst of events is recorded with the transistor simultaneously with a response of the carbon fiber. In that system, the EOS transistors work as ion sensitive devices that record the local pH change that is caused by the vesicles. Whether that approach can be used to record neuronal synapses with their far smaller vesicles has to be studied.

7. Brain Tissue on Chips

Culturing of defined neuronal networks can be avoided when neuronal networks are used that are provided by brains. Brain tissue with a planar structure of the networks is required to attain an efficient interfacing with a planar chip. Organotypic brain slices are particularly promising because they are a few cell layers thick and conserve major neuronal connections. Compared to the culture of dissociated cells, however, there are new problems if we want to interface individual nerve cells in a tissue: (i) We cannot take it for granted that individual nerve cells are attached to individual stimulation capacitors and sensor transistors with a distance of 50 nm. (ii) There are so many neuronal cell bodies embedded in glia cells and in a web of dendrites and axons, that we do not know which cell is coupled to a particular stimulation or recording site. It will be an important task to investigate the structural and electrical features of the tissue-chip contact with similar optical methods as used for cultured cells.

In a first approach, we attempt to achieve an electrical interfacing of organotypic brain slices on the level of neuronal groups, i.e. we stimulate a small area of presynaptic brain tissue with an EOS capacitor and detect the average postsynaptic activity in another small area of tissue with an EOS transistor [16].

7.1. *Two-Way Interfacing*

Fig. 10 shows a classical picture of wiring in a transverse section of rat hippocampus. We place a slice on a chip with an array of capacitors and an array of transistors as shown in Fig. 10. The capacitors are in the CA3 region whereas the transistors are in the CA1 region of the hippocampus. These two areas are connected by axons (Schaffer collateral). When the CA3 regions is stimulated, excitatory synapses in the CA1 regions are activated.

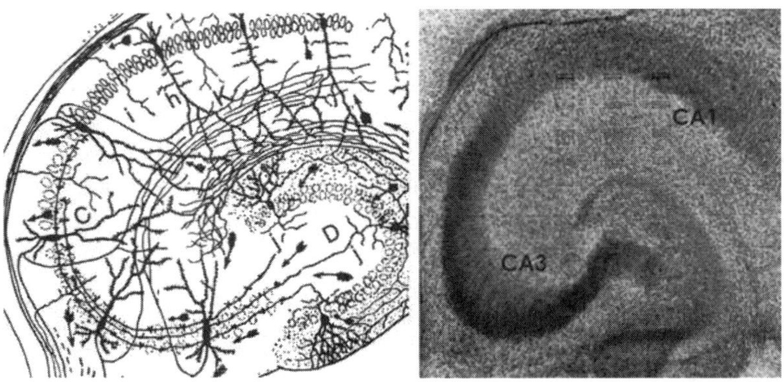

Figure 10. Rat hippocampus on silicon chip. Left: Transverse section of hippocampus that shows the neuronal wiring (Ramon y Cajal, Golgi stain). Right: Hippocampus slice (Nissl stain) cultured on a chip with an array of stimulation EOS capacitors (circles) and an array of EOS transistor (bars).

In fact we are able to stimulate the cell bodies of CA3 with sharp falling voltage ramps applied to a capacitor and to record with a certain delay extracellular voltages in the the region of CA1.The postsynaptic signal appears beyond a threshold of the capacitive stimulus when neurons are excited above the center of the capacitor [16]. An enhanced stimulus leads to an expansion of the area where presynaptic neurons are excited such that more postsynaptic neurons are activated with increasing amplitude of the postsynaptic signal.

7.2. *Mapping with CMOS Chip*

Interfacing of brain-tissue with cellular resolution requires a high density of stimulation and recording sites. In a first approach, the mapping of postsynaptic activity is studied on the level of neuronal groups using a CMOS chip with 16384 sensor transistors on one squaremillimeter. Fig. 11 shows a chip with an organotypic hippocampus slice where the recording array covers the CA3 region. A tungsten electrode is used to stimulate axons that enter the CA3 region (mossy fibers). Fig 11also shows the distribution of extracellular voltage 5 ms after stimulation [17]. We see a negative signal in the region of the dendrites (stratum radiatum) and a positive signal in the region of the cell bodies (stratum pyramidale). The action potential in the presynaptic axons activate the synapses in the dendritic region where current flows into the cells such that a negative extracellular voltage is created. The compensating outward current leads to positive voltage in the region of the cell bodies. The CMOS chip provides a time resolved map of activity with a resolution of 2 kHz.

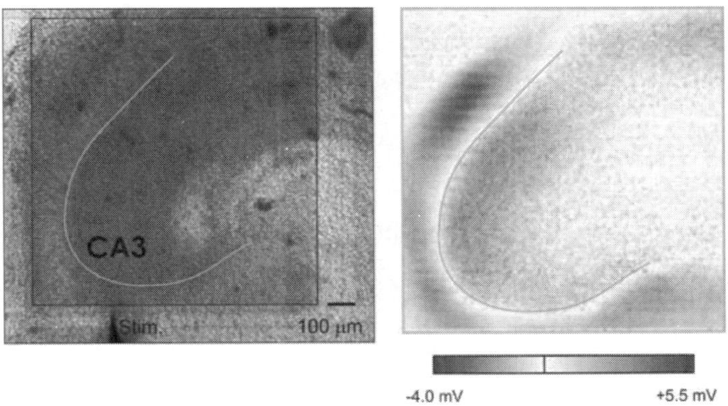

Figure 11. Mapping of brain slice activity by the transistor array of a CMOS chip. Left: Cultured hippocampus slice. The square marks the area of 16384 sensor sites on one squaremillimeter. The slice is stimulated by a tungsten electrode at the bottom (Stim). The gray line is redrawn from the activity map on the right hand side.. Right: Extracellular voltage 5 ms after stimulation: activity of synapses in the area of inward current (within the bent structure) and compensating outward current (periphery of bent structure). The grey line separates the areas of positive and negative voltages.

These experiments are first steps towards a complete interfacing of brain tissue on the level of individual neurons in both directions. Given the features of the hippocampus, we can envisage far reaching experiments on an associative neuronal memory that is integrated with a semiconductor chip.

8. Summary and Outlook

The basic problems on the electrical interfacing of individual nerve cells and semiconductor chips are solved. Important progress was achieved by the development of stimulation EOS capacitors with high dielectric constant, of sensor EOS transistors with low noise and of CMOS chips with a high density of sensor sites. However, with respect to hybrid systems of neuronal networks and digital microelectronics, we are still in an elementary stage. Several directions may be considered: (i) Small defined networks of neurons from invertebrates and mammals are created with defined topology and with learning synapses. (ii) Large neuronal nets are grown on closely packed arrays of two-way contacts fabricated by CMOS technology. (iii) The interfacing of brain slices in both directions is established on the level of neuronal groups with large two-way arrays on CMOS chips. (iv) The structure of the chips is modified such that they can be applied to brains. Of course, visionary dreams of bioelectronic neurocomputers and microelectronic neuroprostheses are unavoidable and exciting. However, they should not obscure the numerous practical problems that must be solved.

Acknowledgments

The work was supported by the Max Planck Society, the Federal Minister of Research, the German Science Foundation (SFB 563) and the European Union (NaChip project).

References

1. P. Fromherz, in: *Bioelectronics.* Eds. I. Willner & E. Katz (Wiley-VCH, Weinheim, 2005) 339.
2. A. Lambacher, and P. Fromherz, *Appl. Phys. A* **63**, 207 (1996).
3. D. Braun, and P. Fromherz, *Biophys. J.* **97**, 1351 (2004).
4. F. Wallrapp, and P. Fromherz, *J. Appl. Phys.* **99**, 114103 (2006).
5. M. Ulbrich, and P. Fromherz, *Appl. Phys. A* **81**, 887 (2005).
6. M. Schmidtner, and P. Fromherz, *Biophys. J.* **90**, 183 (2006).
7. I. Schoen, and P. Fromherz, *Biophys. J.* **92**, 1096 (2007).
8. R. Schätzthauer, and P. Fromherz, *Eur. J. Neurosci.* **10**, 1956 (1998).
9. A. Lambacher, M. Jenkner, M. Merz, B. Eversmann, R.A. Kaul, F. Hofmann, R. Thewes, and P. Fromherz, *Appl. Phys. A* **79**, 1607 (2004).
10. M. Voelker, and P.Fromherz, *Phys. Rev. Lett.* **96**, 228102 (2006).
11. P. Bonifazi, and P. Fromherz. *Adv. Mat.* **14**, 1190 (2002).
12. R.A. Kaul, N.I. Syed, and P. Fromherz. *Phys. Rev. Lett.* **92**, 038102 (2004).
13. G. Zeck, and P. Fromherz, *Proc. Natl. Acad. Sci. USA* **98**, 10457 (2001).
14. M. Merz, and P. Fromherz, *Adv. Funct. Mat.* **15**, 739 (2005).
15. J. Lichtenberger, and P. Fromherz, *Biophys. J.* **92**, 2262 (2007).

16. M. Hutzler, and P. Fromherz, *Eur. J. Neurosci.* **19**, 2231 (2004).
17. M. Hutzler, A. Lambacher, B. Eversmann, M. Jenkner, R. Thewes, and P. Fromherz, *J. Neurophysiol.* **96**, 1638 (2006).

NMR implementation of exponential sums for integer factorization

M. Štefaňák[(1,2)], W. Merkel[(1)], M. Mehring[(3)] and W. P. Schleich[(1)]

[(1)] *Institut für Quantenphysik, Universität Ulm, D-89081 Ulm, Germany*

[(2)] *Department of Physics, FJFI ČVUT v Praze, 115 19 Praha 1 - Staré Město, Czech Republic*

[(3)] *Physikalisches Institut, Universität Stuttgart, D-70550 Stuttgart, Germany*

We briefly summarize the use of Gauss sums to factor numbers and propose an NMR realization of exponential sums.

Keywords: NMR; Gauss sums; integer factorization.

1. Introduction

In 1982 Paul A. Benioff[1] constructed a model for computation using the tools of quantum mechanics. The universal quantum simulator[2] was the motivation of Richard P. Feynman's interest in a quantum computer. However, David Deutsch was one of the first to formulate an algorithm[3] which would benefit from a quantum computer. Nevertheless, the field had to wait almost ten years till the suggestion for factorization with the help of a quantum computer[4] and almost another ten years for the experimental realization[5]. By now quantum information[6–8] is a mature branch of science.

Gauss sums play an important role in number theory[9,10] especially in the context of integer factorization. Recently experimental schemes for integer factorization based on Gauss sums have been proposed[11–17]. Triggered by this work several experiments based on NMR[18,19], cold atoms[20] and ultra-short pulses[21] have successfully demonstrated the possibility to factorize up to 8-digit numbers. In order to save experimental resources we have recently studied the use of truncated exponential sums for integer factorization[22]. In the present paper we propose an experimental implementation of exponential sums following NMR techniques[18].

Our paper is organized as follows. In Section 2 we summarize the central idea of factorizing numbers with Gauss sums. Motivated by the question of resources we then turn in Section 3 to exponential sums. Finally, in Section 4 we propose an NMR method to implement such exponential sums. We conclude in Section 5 by presenting a brief summary and an outlook.

2. Factorization with Gauss sums: the central idea

Our factorization algorithm relies on the truncated Gauss sum

$$\mathcal{A}_N^{(M)}(\ell) \equiv \frac{1}{M+1} \sum_{m=0}^{M} \exp\left[2\pi i\, m^2 \frac{N}{\ell}\right] \tag{1}$$

with $M+1$ terms. Here N is the number to be factorized and ℓ is a trial factor which scans through all integers between 1 and $\lfloor\sqrt{N}\rfloor$.

The capability of the Gauss sum $\mathcal{A}_N^{(M)}$ to factor numbers stems from the fact that for $\ell = q$ where q is an integer factor of $N = q \cdot r$ all phases in $\mathcal{A}_N^{(M)}(q)$ are integer multiples of 2π. Consequently, the terms add up constructively and yield $\mathcal{A}_N^{(M)}(q) = 1$. When ℓ is not a factor the quadratic phases oscillate rapidly with m and $|\mathcal{A}_N^{(M)}(\ell)|$ takes on small values.

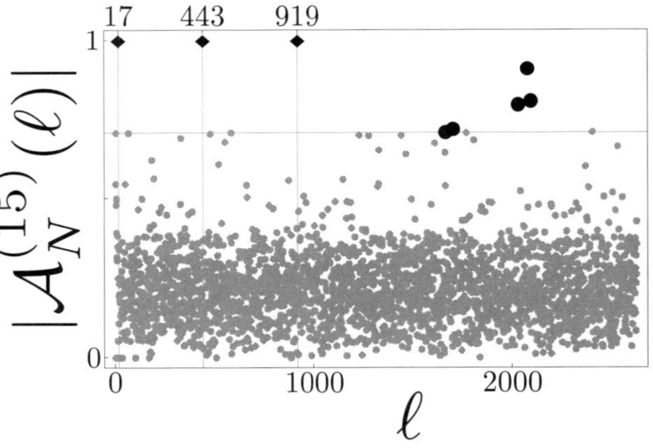

Fig. 1. Factorization interference pattern of the number $N = 6920989 = 17 \cdot 443 \cdot 919$ obtained from the Gauss sum $\mathcal{A}_N^{(M)}$ defined by (1). Here we have chosen the truncation parameter $M \approx \ln N \approx 15$. The factors of N, depicted by diamonds, correspond to the signal value of unity. For most of the non-factors, indicated by gray dots, the signal value is well suppressed. However, for a few non-factors, shown by black dots, the signal is close to that of a factor. Such arguments can be misinterpreted as factors of N. Hence, we call them *ghost factors*. The presence of ghost factors in the factorization interference pattern indicates that the choice $M \approx \ln N$ of the truncation parameter is not appropriate.

In order to factor a number N we analyze the *signal*, i.e. the absolute value $|\mathcal{A}_N^{(M)}(\ell)|$ of the Gauss sum, for integers $1 \leq \ell \leq \lfloor\sqrt{N}\rfloor$. We refer to the graphical representation of this set of data as *factorization interference pattern.*

In Fig. 1 we show an example of such a factorization interference pattern for the number $N = 6920989 = 17 \cdot 443 \cdot 919$ with the choice $M = 15 \approx \ln N$ of the truncation parameter. We find that the factors of N lead to the maximal signal of unity, whereas for most of the non-factors the signal is well suppressed. However, for some non-factors, the so-called *ghost factors*, the signal is still close to that of a factor. Hence, the ghost factors spoil the discrimination of factors from non-factors. Fortunately, we can reduce the signal of a ghost factor by increasing the truncation parameter M. Indeed, we need[23] $M \sim \sqrt[4]{N}$ terms in the sum (1) in order to push the signal of all ghost factors below the threshold[23] of $1/\sqrt{2}$.

3. Exponential sums

The truncation parameter M translates directly into experimental resources. Indeed, all experimental realizations[18–21] of Gauss sums identify M with the number of pulses applied onto the system, or the number of interfering light fields. In order to save resources we have recently studied[22] the exponential sum

$$\mathcal{A}_N^{(M,j)}(\ell) \equiv \frac{1}{M+1} \sum_{m=0}^{M} \exp\left[2\pi i\, m^j \frac{N}{\ell}\right]. \tag{2}$$

In the case of $j = 1$ the exponential sum reduces to a Fourier sum. For $j = 2$ we find the truncated Gauss sum (1), i.e. $\mathcal{A}_N^{(M,2)} \equiv \mathcal{A}_N^{(M)}$. Since the mathematician Ernst Kummer (1810-1893) had studied exponential sums with $j = 3$ this class of exponential sums is called *Kummer sum.*

Exponential sums are capable of discriminating non-factors in exactly the same way as the truncated Gauss sum defined by (1). However, the number M of terms which have to be summed in order to free the signal of all ghost factors of N now scales like[22] $M \sim \sqrt[2j]{N}$ and thus leads to more effective scaling laws.

We illustrate this feature in Fig. 2 where we show the factorization interference patterns of the number $N = 6920989 = 17 \cdot 443 \cdot 919$ resulting from a Fourier sum (left) and a Kummer sum (right). In order to compare with Fig. 1 we keep the truncation parameter $M \approx \ln N \approx 15$. Clearly, the cubic phase varies rapidly enough to suppress the ghost factors which cover almost the complete interference pattern of a Fourier sum.

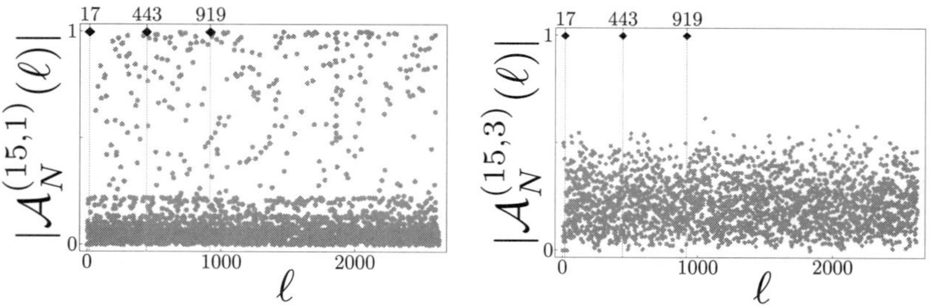

Fig. 2. Effectiveness of factorization with exponential sums illustrated by a Fourier sum (left) and a Kummer sum (right) where the phases are linear and cubic in the summation index m, respectively. In order to compare with Fig. 1 we keep the same number to be factored $N = 6920989 = 17 \cdot 443 \cdot 919$ and the same truncation parameter $M \approx \ln N \approx 15$. Since the linear phase grows slowly the contrast provided by the Fourier sum for $M = 15$ is not sufficient to reveal the factors of N. Indeed, the factorization interference pattern is overloaded with ghost factors. In contrast, the cubic phase varies rapidly and the truncation parameter $M = 15$ is sufficient to suppress all ghost factors.

4. NMR realization

We now turn to the experimental implementation of an arbitrary exponential sum. Here we propose a slightly generalized method used[18] already successfully in the NMR realization of Gauss sums.

We subject an ensemble of spins $I = 1/2$ to a specific sequence of RF pulses as shown in Fig. 3. They lead to a sequence of signals which when summed represent a sum closely related to (2).

The pulse sequence consists of an initiating $\pi/2$-pulse in y-direction which creates the initial density matrix $\rho_{\text{in}} \equiv \mathbb{1}/2 - \epsilon\, I_x$ where $0 < \epsilon < 1$. This pulse is followed by a series of $M+1$ π-pulses with separation 2τ which are individually phase shifted[24,25] with respect to the x-axis of the rotating frame by an angle $\phi_k^{(j)}$. In the laboratory system the corresponding Hamiltonian reads

$$H(t) = \hbar\omega_0\, I_z + 2\pi\hbar \sum_{k=0}^{M} \delta(t-\tau_k)\cos(\omega\tau_k - \phi_k^{(j)})\, I_x \tag{3}$$

where ω_0 and ω denote the frequencies of the transition and the driving field, respectively. The pulses act at times $\tau_k \equiv (2k+1)\tau$ and the phases $\phi_k^{(j)}$ needed to obtain a particular type of exponential sum $\mathcal{A}_N^{(M,j)}$ will be specified later.

With the help of the identity

$$U_n(\alpha) \equiv \exp(-i\alpha I_n) = \cos(\alpha/2)\,\mathbb{1} - 2i\,\sin(\alpha/2)\, I_n \tag{4}$$

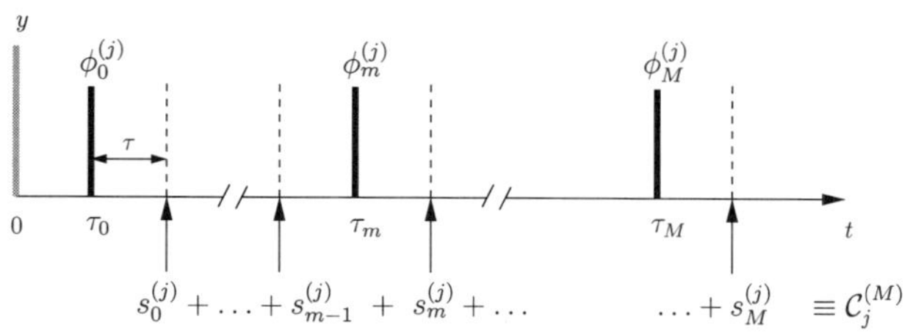

Fig. 3. NMR implementation of exponential sums. The y-pulse which prepares the initial density matrix ρ_{in} is followed after a time τ by a pulse which imprints the phase $\phi_0^{(j)}$ on the spins. This pulse is the first of a sequence of $M+1$ pulses each imprinting a different phase $\phi_m^{(j)}$. At times $\tau_m + \tau \equiv (2m+1)\tau + \tau$ we measure the echo, that is the polarization $s_m^{(j)}$ in x-direction and sum the echoes $s_m^{(j)}$ over all m to obtain $C_j^{(M)}$. In order to implement the exponential sum (2) the phases $\phi_m^{(j)}$ need to be proportional to the number N to be factored and the difference $m^j - (m-1)^j$ since the spin dynamics expressed by $s_m^{(j)}$ depends on the sum over all phases of the previous pulses.

the Hamiltonian $\widetilde{H}$ within the rotating wave approximation, that is in the frame rotating with the frequency $\Delta\omega \equiv \omega_0 - \omega$ takes the form

$$\widetilde{H} = \hbar\Delta\omega\, I_z + \pi\hbar \sum_{k=0}^{M} \delta(t-\tau_k)\left(\cos\phi_k^{(j)}\, I_x + \sin\phi_k^{(j)}\, I_y\right) \tag{5}$$

and the resulting time evolution operator

$$\mathcal{U}_k \equiv U_z(\Delta\omega\tau)U_z(\phi_k^{(j)})U_x(\pi)U_z^{\dagger}(\phi_k^{(j)})U_z(\Delta\omega\tau) \tag{6}$$

of the k-th cycle reflects the three stages shown in Fig. 3: (*i*) Free time evolution for the time τ followed by (*ii*) a π-pulse which imprints the phase $\phi_k^{(j)}$, and (*iii*) another free time evolution during the time τ. The π-pulse eliminates the effects of the free time evolution and ensures that the decoherence due to an inhomogeneous distribution of local fields is compensated by refocussing.

At times $\tau_m + \tau$ we measure the polarization

$$s_m^{(j)} \equiv \frac{\text{Tr}\,(I_x\rho_m)}{\text{Tr}\,(I_x\rho_{\text{in}})} \tag{7}$$

in x-direction with

$$\rho_m \equiv \rho(\tau_m + \tau) = \mathcal{U}_m\rho_{\text{in}}\mathcal{U}_m^{\dagger} \tag{8}$$

where

$$\mathcal{U}_m \equiv \mathcal{U}(\tau_m + \tau) = \prod_{k=0}^{m} \mathcal{U}_k \tag{9}$$

is the product of the time evolution operators $\mathcal{U}_k$ due to all previous pulses.

When we apply (4) to (6) we find

$$\mathcal{U}_k = (-i) \begin{pmatrix} 0 & e^{-i\phi_k^{(j)}} \\ e^{i\phi_k^{(j)}} & 0 \end{pmatrix} \tag{10}$$

which yields for the average $\mathcal{C}_j^{(M)}$ over the signals $s_m^{(j)}$ given by (7) the expression

$$\mathcal{C}_j^{(M)} \equiv \frac{1}{M+1} \sum_{m=0}^{M} s_m^{(j)} = \frac{1}{M+1} \sum_{m=0}^{M} \cos\left(\sum_{k=0}^{m} (-1)^k 2\phi_k^{(j)} \right). \tag{11}$$

Hence, the spin dynamics expressed by the signal $s_m^{(j)}$ depends on its complete history, that is the phases of all previous pulses.

The sum $\mathcal{C}_j^{(M)}$ is closely related to the exponential sum (2). Indeed, for the particular choice

$$\phi_k^{(j)} \equiv \begin{cases} (-1)^k (k^j - (k-1)^j)\pi \frac{N}{\ell} & \text{for } k \geq 1 \\ 0 & \text{for } k = 0 \end{cases}, \tag{12}$$

of the phases $\phi_k^{(j)}$ the average signal turns into

$$\mathcal{C}_j^{(M)} = \frac{1}{M+1} \sum_{m=0}^{M} \cos\left(2\pi m^j \frac{N}{l} \right), \tag{13}$$

which is equal to the real part of the exponential sum (2).

5. Conclusions

In the present paper we have briefly summarized our work on factoring numbers with Gauss and exponential sums. Three results stand out most clearly: (*i*) the maxima of exponential sums at integers yield factors, (*ii*) in order to save resources it pays-off to use exponential sums whose contributing phase factors oscillate more rapidly than the ones of Gauss sums, (*iii*) we have proposed an experimental scheme to realize an arbitrary exponential sum.

Our approach towards factorization rests solely on interference and is very much in the spirit of recent critical discussions[26,27] on the connection between quantum mechanics and factoring. Indeed, any classical system displaying interference could be used to implement exponential sums. For this reason there is no speed-up compared to a classical computer and the resources necessary to factor a number N scale exponentially.

Many ideas how to combine factorization based on exponential sums with entanglement and to develop a Shor-like algorithm offer themselves. These conclusions are neither the time nor the space to elaborate in detail on these new goals. For this reason we only mention three different approaches: (*i*) Let two harmonic oscillators interact with a two-level atom such that the oscillators pick-up quadratic phases of different signs. Make joint measurements on the atom and one oscillator. The resulting state is described by the real or the imaginary part of a Gauss sum. (*ii*) Entangle two harmonic oscillators. Whereas the register is in a number state the coherent state has an amplitude determined by an exponential sum. (*iii*) Use cold atoms in an optical lattice described[28] by the Bose-Hubbard model. Due to their interaction the occupation number appears quadratically giving rise to quadratic phases as observed experimentally[29] by the collapse and the revival of a Bose-Einstein condensate.

Presently we are in the process of developing these ideas further. Manuscripts devoted to these topics are in preparation.

6. Acknowledgement

We thank D. Haase and M. S. Zubairy for many stimulating discussions. Moreover, W. P. S. enjoyed the exciting *International Symposium on Contemporary Physics* and thanks the organizers for the most inspiring meeting. His participation in Islamabad was also made possible by the Alexander von Humboldt-Stiftung. W. M. and W. P. S. acknowledge financial support from the Ministry of Science, Research and the Arts of Baden-Württemberg in the framework of the Quantum Information Highway A8 and the Center of Quantum Engineering. M. Š. is grateful for a stipend from the European Community in the framework of CONQUEST. Moreover, W. P. S. also would like to thank the Alexander von Humboldt-Stiftung and the Max-Planck-Gesellschaft for receiving the Max-Planck-Forschungspreis.

References

1. P. A. Benioff, *Int. J. Theor. Phys.* **21**, 177 (1982)
2. R. P. Feynman, *Int. J. Theor. Phys.* **21**, 467 (1982)
3. D. Deutsch, *Proc. R. Soc. Lond.* A **400**, 97 (1985)
4. P. Shor, in: *Proceedings of the 35th Annual Symposium on Foundations of Computer Science* (IEEE Computer Society, New York, 1994)
5. L. M. K. Vandersypen, M. Steffen, G. Breyta, C. S. Yannoni, M. H. Sherwood and I. L. Chuang, *Nature* (London) **414**, 883 (2001)
6. S. Stenholm and K.-A. Suominen, *Quantum Approach to Informatics* (John-Wiley, New York, 2005)

7. M. A. Nielsen and I. L. Chuang, *Quantum Computation and Quantum Information* (Cambridge University Press, Cambridge, 2000)
8. W. P. Schleich and H. Walther, *Elements of Quantum Information* (Wiley-VCH, New York, 2007)
9. H. Davenport, *Multiplicative Number Theory* (Springer, New York, 1980).
10. H. Maier and W. P. Schleich, *Prime Numbers 101: A Primer on Number Theory* (Wiley-VCH, New York, 2007)
11. J. F. Clauser and J. P. Dowling, *Phys. Rev.* A **53**, 4587 (1996)
12. W. G. Harter, *Phys. Rev.* A **64**, 012312 (2001)
13. H. Mack, M. Bienert, F. Haug, M. Freyberger and W. P. Schleich, *Phys. Stat. Sol.* (b) **233**, 408 (2002);
14. H. Mack, M. Bienert, F. Haug, F. S. Straub, M. Freyberger and W. P. Schleich, in: *Experimental Quantum Computation*, Eds. P. Mataloni and F. De Martini, (Elsevier, Amsterdam, 2002)
15. W. Merkel, O. Crasser, F. Haug, E. Lutz, H. Mack, M. Freyberger, W. P. Schleich, I. Sh. Averbukh, M. Bienert, B. Girard, H. Maier and G. G. Paulus, *Int. J. of Mod. Phys.* B **20**, 1893 (2006)
16. W. Merkel, I. Sh. Averbukh, B. Girard, G. G. Paulus and W. P. Schleich, *Fortschritte der Physik* **54**, 856 (2006)
17. W. Merkel, I. Sh. Averbukh, B. Girard and W. P. Schleich, *Phys. Rev.* A (2007), to be published
18. M. Mehring, K. Müller, I. Sh. Averbukh, W. Merkel and W. P. Schleich, *Phys. Rev. Lett.* **98**, 120502 (2007)
19. T. S. Mahesh, N. Rajendran, X. Peng and D. Suter, *Phys. Rev.* A **75**, 062303 (2007)
20. M. Gilowski, T. Wendrich, T. Müller, Ch. Jentsch, W. Ertmer, E. M. Rasel and W. P. Schleich, submitted to *Phys. Rev. Lett.* (2007)
21. D. Bigourd, B. Chatel, W. P. Schleich and B. Girard, submitted to *Phys. Rev. Lett.* (2007)
22. M. Štefaňák, D. Haase, W. Merkel, M. S. Zubairy and W. P. Schleich, in preparation for *J. Phys.* A (2007)
23. M. Štefaňák, W. Merkel, W. P. Schleich, D. Haase and H. Maier, submitted to *New J. Phys.* (2007)
24. W. S. Warren, D. P. Weitekamp and A. Pines, *J. Magn. Reson.* **40**, 581 (1980)
25. R. R. Ernst and G. Bodenhausen and A. Wokaun, *Principles of Nuclear Magnetic Resonance in One and Two Dimensions*, (Clarendon Press, Oxford, 1987)
26. N. D. Mermin, *Physics Today* **60**, 8 (2007)
27. M. S. Zubairy, *Science* **316**, 554 (2007)
28. D. Jaksch, C. Bruder, J. I. Cirac, C. W. Gardiner and P. Zoller, *Phys. Rev. Lett.* **81**, 3108 (1998)
29. M. Greiner, O. Mandel, T. W. Hänsch and I. Bloch, *Nature* **419**, 51 (2002)

SECTION III

NONLINEAR PHENOMENA AND PLASMA PHYSICS

Complexity and hydrodynamic turbulence

K.R. Sreenivasan

The Abdus Salam International Centre for Theoretical Physics
Strada Costiera 11, 34014 Trieste, Italy
E-mail: krs@ictp.it

Physics is a seamless dialogue with Nature that transcends the divisions made by practising physicists—often for convenience, often imposed by their own limitations. It is difficult to sustain the notion that one branch of physics can satisfactorily answer all the questions of Nature. I expand on this theme briefly, and also technically, using the example of hydrodynamic turbulence.

Keywords: Complexity; Hydrodynamic turbulence.

1. Introduction

At the Symposium I reported the properties of cryogenic turbulence, both above and below the lambda-point. Much of that work has been published[1], so it is hard to justify another account of the same work. Since it is not especially useful to write for these proceedings a paper on specialized developments that have occurred since then on the subject of my talk, I have chosen to set down a few general remarks on the physics that is close to my interests, and add a few technical details to justify them.

Physics is about intelligent and rigorous dialogue with Nature; the aim is to discover exact relationships between measurable quantities, and the preferred language is mathematical. Agreement with experiments is essential but not in the way a lay person might imagine. Newton's first law, for example, is evidently not satisfied in our everyday experience (bodies set in uniform motion do not stay in uniform motion forever), but one can seek the confirmation of the law in controlled circumstances. By successively decreasing the friction between a moving body and its surroundings, one can presumably observe the body to remain in motion for longer and longer periods of time, leading us to infer that Newton's first law is correct in the absence of friction.

Within the broad definition of physics mentioned above, its different branches may have different characteristic approaches. One common element, however, is the goal of discovering general laws that underlie Nature's myriad manifestations. Questions about these general laws have led 20th century physics to inquire into the structure of increasingly smaller constituents of matter[2], culminating in modern particle physics and its more recent developments such as superstrings.

The reductionist approach assumes that the world around us can be understood in terms of the properties of a few small building blocks (particles or strings), and that everything can be understood from them, or the "first principles". The approach has worked well in some instances—for example in characterizing the structure of the hydrogen atom from quantum mechanics—but this "first principles" understanding is not quantitative for anything but the simplest of problems. No one has yet deduced the structure of macromolecules using the principles that were successful with respect to the hydrogen atom. In general, one cannot connect hierarchies of physical description without some empirical input; no effort has been made, for instance, to compute the structure of organic molecules from particle physics, and to go from these molecules to cells and living organisms. The question is not simply one of logistics but one of principle: it is not clear, even in principle, if one can understand the collective motion of a large number of nonlinearly interacting scales from this bottom-up approach. Perhaps more provocatively, it is not even clear if the knowledge organized at one level is essential for a deep understanding of the knowledge at a distantly coarser level.

Are there different fundamental laws that are applicable at different levels of description, with broad generality of their own? For instance, is self-organization, which is characteristic of living systems (among others), subject to some fundamental laws that one cannot deduce even *in principle* from particle physics?

A large body of scientists today thinks that reductionism, the cornerstone of much of the 20th century physics, has limitations of principle despite its enormous successes; that a deductive link does not exist between the finest constituents of matter and phenomena that occur on the human scale; that one needs an equally deep understanding of the so-called emergent phenomena regulated by different organizing principles ("the psychology of the lynch mob vanishes when you interview individual participants"[3]); that these organizing principles are equally deep in content and structure; that a knowledge of the constituent parts, however complete, cannot describe the whole. Perhaps it is too much to say that physics at the turn of the 21st century is undergoing a crisis similar to that at the turn of the last, but physics is changing its landscape.

As mentioned already, physics looks for generality of laws. The question is whether there are any such laws at all that describe a broad class of systems with many nonlinearly interacting degrees of freedom. Here, I pose the question at the quantitative level of rigor, not simply at a qualitative one; for instance, it is not enough that daily variations in financial markets *look* closely similar to gradients of the velocity field in a turbulent flow. If quantitative laws do cut across seemingly unrelated problems, in what form will they appear? Since it is known that nonlinear systems cannot be predicted except for short periods of time, what form should comparisons with experiment take? These questions are not new; equilibrium statistical mechanics faced them in a certain way more than 100 years ago. The domain of our consideration now is much broader and the urgency for the answer far greater, because a multitude of problems in Nature that spill out of the reductionist box are known not to obey the laws of equilibrium statistical mechanics.

2. The case for hydrodynamic turbulence

If we are interested in discovering laws underlying systems with many strongly interacting degrees of freedom and are far from equilibrium, it is important to begin with a study a few of them with the same rigor and control for which particle physics, say, is well known. Here, I make the case that hydrodynamic turbulence, which arises in flowing fluids, is an ideal paradigm. My first point is that the dynamical equations for the motion of fluids are known to great accuracy, which means that understanding their analytic structure can greatly supplement experimental queries; in just the same way, computer simulations—even if they require much investment of time and money—can be far more useful here than for many other problems of the condensed phase, in which the interaction potential among microscopic parts is often simply an educated guess. When driven hard, fluids develop irregular motion in time and space—this being the state of turbulence. The stochasticity of turbulence (and of all systems that are driven hard) means that one may discern only laws that concern statistical behavior. If we are fortunate, these laws are universal in some well-understood sense. This is the aspect I wish to address briefly.

To be specific, consider a jet of fluid emerging from an orifice. It is immaterial if the fluid is air (emerging into air medium), water (emerging into water medium) or mercury (emerging into mercury medium)—already a great generalization. Let the jet velocity be V, the orifice diameter D and fluid viscosity ν. The Reynolds numbers $Re = VD/\nu$ should be "high enough" for our considerations to hold. The requirement of high Reynolds number is no more constraining than one on the energy levels needed to study elementary particles. Some forcing is essential to maintain the flow steady, else it will grind to a halt eventually because of dissipation. The forcing in this instance is provided by the momentum at the jet orifice; typically the scale of forcing is of the order D, but not exactly so. Let us call it L.

Further from the orifice there appear a range of motions much smaller that L juxtaposed randomly in space and time. These scales are produced by the nonlinearity of the equations of motion. It is believed that all important properties of turbulence are contained in the Navier-Stokes equations for the fluid motion which can be written for unit density as

$$\mathbf{v}_t + (\mathbf{v} \cdot \nabla)\mathbf{v} = -\nabla p + \mathbf{f} + \nu \Delta \mathbf{v}\,, \qquad (1)$$

supplemented by the incompressibility condition $\nabla \cdot \mathbf{v} = 0$. Here $\mathbf{v}$ is the velocity vector, p is the local pressure, f is the forcing and the suffix t stands for time-derivative. In (1), f is a stochastic forcing term that suitably abstracts the mechanism for sustaining turbulence.

The production of smaller and smaller scales is supposed to proceed in a cascade-like process[4]. The process is operational until viscous dissipation truncates it at a characteristic scale η (which is proportional to the 3/4-power of ν) that is much smaller than L. The scales between L and η, which remain unaffected directly by forcing as well as viscosity, or by L as well as η, form the so-called inertial range. The kinetic energy injected at scales of the order L is transferred through the inertial range and dissipated eventually at scales of the order η. Indeed, nonlinear energy transfer process is a dominant feature in the inertial range.

Because neither L nor η is relevant in the inertial range, it is natural to ask if some attributes of the inertial range are universal. Another question of the universality problem concerns features that are common to different non-equilibrium systems. The quest for universality is motivated by the hope of identifying some general principles that govern far-from-equilibrium systems, just as variational principles govern thermal equilibrium. As already mentioned, the dynamical features of turbulence are irregular, and so the questions can be posed and answered only in terms of statistical averages—computed for appropriate regions of space, intervals of time, or suitably defined ensembles.

I wish to emphasize that there are many other interesting and important problems in turbulence, especially when linked to practical applications.

3. The nature of the inertial range

The universality of the inertial-range is studied by constructing the so-called structure functions, which are moments of velocity differences δv_r across a scale of size r. Structure functions of order n are defined as $S_n \equiv \left\langle [(\mathbf{v}(\mathbf{r},t) - \mathbf{v}(\mathbf{0},t)) \cdot \mathbf{r}/r]^n \right\rangle \equiv \left\langle (\delta v_r)^n \right\rangle$, with $r \equiv |\mathbf{r}|$. The angular brackets indicate a suitable average. Generalization of structure functions, which we will not consider here, are statistical correlations of velocity differences at the vertices of more complex configurations such as triangles, quadrilaterals, and so forth. The study of correlators of different types is thus a study of different geometrical configurations.

The third-order structure function in the inertial range of scales can be related to the energy flux by an exact flux-constancy relation derived by Kolmogorov[5] from (1):

$$S_3 \equiv -\frac{4}{5}\langle\epsilon\rangle r\ . \tag{2}$$

This is known as the 4/5-ths law. Here $\langle\epsilon\rangle = \nu\langle|\nabla\mathbf{v}|^2\rangle$ is the average rate of energy dissipation per unit mass that can be identified with the energy flux over scales.

A few comments are in order. The above result for the inertial range is a direct consequence of the dynamical equations without any hypothesis. The only requirement is that the Reynolds number must be "sufficiently high"; further increase of the Reynolds number merely extends the inertial range without altering its scaling. Lest this result be thought of as merely mathematical, it should be stressed that measurements in real flows do indeed obey the 4/5-ths law; see, e.g.,[6]. The interesting point is that the law is valid for turbulence in all fluids, no matter what their molecular structure: Water, mercury, nitrogen and liquid helium all follow the same law. The nuclear structure of the molecules or atoms making up the fluid makes no difference. Indeed, one can say that the atomic properties and the behavior of turbulence are disconnected, in so far as we are concerned with this law. It is thus impossible to say that one can work up towards this law from the bottom, except in some generic fashion.

No one has been able to deduce theoretically moments of other orders. Kolmogorov[5] assumed scale invariance and obtained a tidy result. Scale invariance means that the symmetry broken by large-scale forcing is restored when turbulence scales become small compared to the forcing scale, L. Thus, the structure functions are thought to be universal power laws containing neither L nor η. This picture readily yields exponents ζ_n that depend linearly on the moment order, or $S_n \propto r^{\zeta_n}$ with $\zeta_n = n/3$.

4. Dissipation anomaly and anomalous scaling

The modern experimental evidence[7] is that the exponents depart from the scale-invariant result, and the forcing scale L appears explicitly in characterizing the inertial range. This non-restoration of scale invariance even at small scales, is arguably an important feature of turbulence, and sets it apart from the usual critical phenomena[8]: One needs to work out the behavior of each order moment independently without succumbing to dimensional analysis. Anomalous scaling in turbulence is such that $\zeta_{2n} < n\zeta_2$ so that S_{2n}/S_2^n for $n > 2$ increases as $r \to 0$. Relative growth of high moments means that strong fluctuations become more probable as the scales become smaller. One practical consequence is that it limits our ability to produce realistic models for small-scale turbulence.

We have already mentioned that scale invariance yields the explicit expression $\zeta_n = n/3$ for the scaling exponents. However, if scale invariance is violated in fluid turbulence, what else can be used to determine the scaling exponents? Conservation laws impose constraints on the dynamics, and so conserved quantities (or their fluxes) play an essential role in addressing this question. Conservation laws are broken in fluid turbulence by the large-scale forcing (usually through boundary conditions) as well as the dissipation at small scales (usually through fluid viscosity). Indeed, the basic dynamics of the inertial range is not the conservation of energy (whose application gives a drastically unrealistic result of equipartition) but of the energy flux across scales from the large to the small. In the steady state, this flux of energy equals the dissipation of the turbulent kinetic energy at small scales of the order η. For the first sight, it might appear from the definition of $\langle \epsilon \rangle$ below equation (2) that the dissipation rate of turbulent energy would vanish as $\nu \to 0$ (or as $Re \to \infty$), but an important feature of turbulence is that $\langle \epsilon \rangle$ remains finite in this limit: no matter how wide is the scale-range participating in the energy cascade, it carries the same flux.

The finiteness of energy dissipation even in the limit of vanishing viscosity is probably the first example of what is called "anomaly" in modern field-theoretical language: A symmetry of the inviscid equation (here, time-reversal invariance) remains broken even as the symmetry-breaking factor (viscosity) vanishes[9]. Recall from equation (2) that the third-order statistics of turbulent velocity increments are determined completely by $\langle \epsilon \rangle$, whose finiteness is a consequence of the breakdown of time-reversal symmetry in the inertial range. Indeed, velocity changes sign under time reversal so nonzero S_3 means time-irreversibility. Just as conservation of the kinetic energy is broken by viscosity (no matter how small), and this broken symmetry governs the third-order structure function, are there other candidate integrals of motion whose broken symmetries yield other structure functions? This is a fundamental question of turbulence—indeed, in a suitable analogous fashion, of modern statistical physics as a whole.

5. Statistical conservation laws

I shall now briefly discuss laws which are conserved only on the average, yet determine the statistical properties of strongly fluctuating systems. Since it is always possible to find some averages that do not change in random systems, our question should be posed better: Is it possible to find quantities that are expected to change on dimensional grounds yet remain constant because they are governed by some statistical conservation laws? How to build specific functions that are conserved on the average, and how to determine the scaling exponents from them? As fluid particles move in a random flow, an n-particle cloud disperses and grows in size while fluctuations in the shape of the cloud decrease in magnitude. Therefore, one may look for suitable functions of size and shape that are conserved because the growth of distances is compensated by the decrease of shape fluctuations.

For the simplest case of Brownian diffusion, the time derivative of the mean of any function of distances between particles is the Laplacian of this function. Harmonic polynomials turn Laplacian into zero (and can thus be called zero modes of Laplacian), and so are conserved on the average. For Laplacian diffusion, the zero modes are polynomials in R^2 or t so that the scaling dimension of the n-particle mode is $\zeta_n = 2n$. The dependence on n is linear because the particles move independently.

The zero modes governing the decrease of shape fluctuations exist for turbulent diffusion as well. The question of how the statistical conservation laws lead to anomalous scaling has been analyzed quantitatively for a model problem of advection of a passive scalar[10]. A scalar is a quantity such as temperature or concentration of an admixture that is advected by the turbulent velocity field. The scalar is passive if it does not affect the velocity field advecting it. For instance, if the scalar is the temperature, the heating must be small enough so as not to affect the dynamics. The model of Ref. [10] considers the case of a passive scalar field that is advected by a Gaussian velocity field that oscillates infinitely rapidly (the auto-correlation is a delta function); however, its spatial correlation follows the power-law characteristic of the real velocity field. For this model, it is shown that the anomaly arises from the statistical conservation laws of geometric origin, not by the dynamical conservation laws. The statistical conservation laws break the scale invariance in the inertial range and scalar turbulence "knows more" about forcing than just the value of the flux. For a summary of the important developments on this problem, see Ref. [11].

6. Closing remarks

My main point here is that a number of systems of great practical interest operate under conditions that are far from equilibrium and develop seemingly common properties. It is unlikely that these properties can be explained from the finest ingredients of matter. It would be most interesting if these properties possess quantitative universality but this quest has not yet been addressed satisfactorily. The first step in this direction requires that we study at least a few such systems precisely, and understand their structure well. Hydrodynamic turbulence is one of the best examples in this category.

To make the case further, I discussed the anomalous scaling of the inertial range in turbulence. The anomaly arises because the scale invariance is broken. Under such circumstances, the scaling exponents are determined by new types of statistical conservation laws. The solutions of the Kraichnan model[10] have elucidated this point quantitatively. The conserved quantities involve the geometry of multi-point configurations carried by the flow. It is possible that this lesson is of universal validity, i.e. for other nonlinear multiscale systems which possess statistical conservation laws and anomalous scaling exponents.

A postscript may be appropriate. In part through direct and indirect influence of Professor Abdus Salam, the founding director of ICTP, Pakistan has had a few excellent physicists in particle physics and related fields.

However, it has not developed a similarly strong tradition in the science of non-equilibrium and nonlinear physics. It is indeed high time that this development occurred. My hope is that the National Centre for Physics in Islamabad, which hosted this Symposium, will be able to develop and foster such an activity.

Part of this note is based on a previous article with G. Falkovich[12]. I thank Joe Niemela for reading a draft.

References

1. J.J. Niemela and K.R. Sreenivasan, *J. Low Temp. Phys.* **143**, 163-212 (2006); G.P. Bewley, D.P. Lathrop and K.R. Sreenivasan, *Nature* **441**, 588 (2006).
2. A. Pais, *Subtle Is the Lord: The Science and the Life of Albert Einstein*, Oxford University Press, 1982.
3. J. Gleick, *Chaos: Making a New Science*, Viking, New York, 1987.
4. L.F. Richardson, *Proc. R. Soc. Lond. A* **110**, 709-737 (1926).
5. A.N. Kolmogorov, *Dokl. Akad. Nauk. SSSR* **30**, 9-13 (1941); **32**, 16-18 (1941); reproduced in *Proc. Roy. Soc. A* **434**, 9-17 (1991).
6. U. Frisch, *Turbulence: The Legacy of A.N. Kolmogorov*, Cambridge University Press, 1995.
7. K.R. Sreenivasan, *Annu. Rev. Fluid Mech.* **29**, 435-472 (1997).
8. N. Goldenfeld, *Lectures on Phase Trsnsitions and the Renomalization Group*, Addison-Wesley Publishing Co., Reading, MA, 1992.
9. K.R. Sreenivasan, *Phys. Fluids* **27**, 1048-1051 (1984); see also *Phys. Fluids* **10**, 528-529 (1998) and Y. Kaneda, T. Ishihara, K. Itakura and A. Uno, *Phys. Fluids* **15**, L21-L24 (2003). A direct analogy between dissipative anomalies in turbulence and axial anomaly (the breakdown of chirality conservation) in quantum field theory was noticed by A.M. Polyakov, *Nucl. Phys. B* **396**, 367-385 (1993).
10. R.H. Kraichnan, *Phys. Fluids* **11**, 945 (1968).
11. G. Falkovich, K. Gawędzki and M. Vergassola, *Rev. Mod. Phys.* **73**, 913-975 (2001).
12. G. Falkovich and K.R. Sreenivasan, *Phys. Today* **59** (April issue), 43-49 (2006).

Nonlinear interactions in quantum systems

P. K. Shukla* and B. Eliasson

Institut für Theoretische Physik IV, Fakultät für Physik und Astronomie, Ruhr-Universität Bochum, D-44780 Bochum, Germany

** E-mail: ps@tp4.rub.de*

www.tp4.rub.de

We present simulation studies of the formation and dynamics of dark solitons and vortices, and of nonlinear interactions between intense circularly polarized electromagnetic (CPEM) waves and electron plasma oscillations (EPOs) dense in quantum electron plasmas. The electron dynamics in the latter is governed by a pair of equations comprising the nonlinear Schrödinger and Poisson system of equations, which conserves electrons and their momentum and energy. The dynamics of the CPEM waves is also governed by a nonlinear schrödinger equation, which is nonlinearly coupled with the nonlinear Schrödinger equation of the EPOs via the relativistic ponderomotive force, the relativistic electron mass increase in the CPEM field, and the electron density fluctuations. The present governing equations in one spatial dimension admit stationary solutions in the form a dark envelope soliton. The dynamics of the latter reveals its robustness. Furthermore, we numerically demonstrate the existence of cylindrically symmetric two-dimensional quantum electron vortices, which survive during collisions. The nonlinear equations admit the modulational instability of an intense CPEM pump wave against EPOs, leading to the formation and trapping of localized CPEM wave pipes in the electron density hole that is associated with a positive potential distribution in our dense plasma.

Keywords: Quantum plasma; Electromagnetic waves.

1. Introduction

Quantum plasmas are ubiquitous in micro-mechanical systems and ultra-small electronic devices [1], in laser and micro-plasmas [2], and in dense astrophysical environments [3]. In such plasmas, quantum mechanical effects (e.g. tunnelling) are important since the de Broglie length of the charge carriers (e.g. electrons and holes/positrons) is comparable to the dimensions of the system. Although its foundation was laid down by Pines [4], who discussed the properties of electron plasma oscillations (EPOs) in a dense

Fermi plasma, there has recently been a growing interest [5–11] in investigations of relevant quantum hydrodynamical (QHD) models for charged particle systems and collective interactions in quantum plasmas. The QHD model has been introduced to deal with negative differential resistance, and resonant tunnelling phenomena in microelectronic devices [12]. Quantum transport models similar to the QHD plasma model has also been used in superfluidity [13] and superconductivity [14], as well as the study of metal clusters and nanoparticles,where they are referred to as nonstationary Thomas-Fermi models [15].

The high-density, low-temperature quantum Fermi plasma is significantly different from the low-density, high-temperature "classical plasma" obeying the Maxwell-Boltzmann distribution. In a very dense quantum plasma, there are new equations of state [6,18,19] associated with the Fermi-Dirac plasma particle distribution function and there are new quantum forces involving the quantum Bohm potential [12] and the electron-1/2 spin effect [20] due to magnetization. It should be noted that very dense quantum plasmas exist in intense laser-solid density plasma interaction experiments [21–24], in laser-based inertial fusion [25], in astrophysical and cosmological environments [3,26–28], and in quantum diodes [29–31].

It has been recently recognized [10,32,33] that quantum mechanical effects play an important role in intense laser-solid density plasma interaction experiments. In the latter, there are nonlinearities [34] associated with the electron mass increase in the electromagnetic (EM) fields and the modification of the electron number density by the relativistic ponderomotive force. Relativistic nonlinear effects in a classical plasma is very important, because they provide the possibility of the compression and localization of intense electromagnetic waves. In this Letter, we consider nonlinear interactions between intense CPEM waves and EPOs in dense quantum plasmas, which are relevant for a variety of applications in laboratories [23,24].

In this paper, we investigate, by means of computer simulations, the formation and dynamics of dark/gray envelope solitons and vortices in quantum electron plasmas with fixed ion background. The results are relevant for the transport of information at quantum scales in micro-plasmas as well as in micro-mechanical systems and microelectronics. For our purposes, we shall use an effective Schrödinger-Poisson model [5–9], which was developed by employing the Wigner-Poisson phase space formalism on the Vlasov equation coupled with the Poisson equation for the electric potential. Such a model was originally derived by Hartree in the context of atomic physics for studying the self-consistent effect of atomic electrons on the Coulomb potential of the nucleus. We also present theoretical and simulation studies of the CPEM wave modulational instability against EPOs, as well as the trapping of localized CPEM waves into a quantum electron hole in very dense quantum plasmas, which may be relevant for the next generation intense laser-plasma interaction experiments.

2. Dark solitons and vortices in a dense quantum plasma

In this section, we discuss the nonlinear properties and dynamics of dark solitons and vortices in a quantum plasma [19]. Generalizing the one-dimensional Schrödinger-Poisson system of equations [6] to multi-space dimensions, we have

$$i\frac{\partial\Psi}{\partial t} + A\nabla^2\Psi + \varphi\Psi - |\Psi|^4\Psi = 0, \tag{1}$$

and

$$\nabla^2\varphi = |\Psi|^2 - 1, \tag{2}$$

where the wave function Ψ is normalized by $\sqrt{n_0}$, the electrostatic potential φ by T_F/e, the time t by $\hbar T_F$ and the space $\mathbf{r}$ by λ_D. We have introduced the notations $\lambda_D = (\epsilon_0 T_F/n_0 e^2)^{1/2}$ and $A = \Gamma_Q/2$, where the quantum coupling parameter $\Gamma_Q = e^2 m/\hbar\epsilon_0 n_0^{1/3}$ can be both smaller and larger than unity for typical metallic electrons [6]. Here n_0 is the equilibrium electron particle density, $T_F \simeq \hbar n_0^{2/3}/m_e$ is the Fermi temperature (in energy units), m_e is the electron mass, e is the magnitude of the electron charge, ϵ_0 is the electric vacuum permittivity, and $\hbar$ is the Planck constant divided by 2π. Strictly speaking, the nonlinearity $|\Psi|^4$ in the last term in the left-hand side of Eq. (1) was derived for the one-dimensional model [6] and takes the form $|\Psi|^{4/D}$ in D dimensions. However, our numerical investigations of the profiles of dark solitons and vortices have shown very small differences if we use $D = 2$ (for two dimensions) instead of $D = 1$; henceforth, we will keep Eqs. (1) and (2) in the present form. The system (1) and (2) is supplemented by the Maxwell equation

$$\partial\mathbf{E}/\partial t = iA\left(\Psi\nabla\Psi^* - \Psi^*\nabla\Psi\right), \tag{3}$$

where the electric field $\mathbf{E} = -\nabla\varphi$. The system of equations (1)–(3) conserves the number of electrons $N = \int|\Psi|\,d^3x$, the electron momentum $\mathbf{P} = -i\int\Psi^*\nabla\Psi\,d^3x$, the electron angular momentum $\mathbf{L} = -i\int\Psi^*\mathbf{r}\times\nabla\Psi\,d^3x$, and the total energy $\mathcal{E} = \int(-\Psi^*A\nabla^2\Psi + |\nabla\varphi|^2/2 + |\Psi|^6/3)\,d^3x$. We note that one-dimensional version of Eq. (1) without the φ-term has also been used to describe the behaviour of a Bose-Einstein condensate [16].

Let us first consider a quasi-stationary, one-dimensional structure moving with a constant speed v_0, and make the ansatz $\Psi = W(\xi)\exp(iKx - i\Omega t)$, where W is a complex-valued function of the argument $\xi = x - v_0 t$, and K and Ω are a constant wavenumber and frequency shift, respectively. By the choice $K = v_0/2A$, we can then write the coupled system of equations as

$$\frac{d^2W}{d\xi^2} + \lambda W + \frac{\varphi W}{A} - \frac{|W|^4 W}{A} = 0, \tag{4}$$

and

$$\frac{d^2\varphi}{d\xi^2} = |W|^2 - 1, \tag{5}$$

where $\lambda = \Omega/A - v_0^2/4A^2$ is an eigenvalue of the system. From the boundary conditions $|W| = 1$ and $\varphi = 0$ at $|\xi| = \infty$, we determine $\lambda = 1/A$ and $\Omega = 1 + v_0^2/4A$. The system of Eqs. (4) and (5) supports a first integral in the form

$$\begin{aligned} H =& A\left|\frac{dW}{d\xi}\right|^2 - \frac{1}{2}\left(\frac{d\varphi}{d\xi}\right)^2 + |W|^2 \\ &- \frac{|W|^6}{3} + \varphi|W|^2 - \varphi - \frac{2}{3} = 0, \end{aligned} \tag{6}$$

where we have used the boundary conditions $|W| = 1$ and $\varphi = 0$ at $|\xi| = \infty$.

We have solved (4) and (5) numerically and have presented the results in Fig. 1. Here we have plotted the profiles of W^2 and φ for a few values of A, where W was set to -1 on the left boundary and to $+1$ on the right boundary, i.e. the phase shift is 180 degrees between the two boundaries. We see that we have solutions in the form of a dark soliton, with a localized depletion of the electron density $N_e = |W|^2$, and where W has different sign on different sides of the solitary structure. The local depletion of the electron density is associated with a positive potential. Larger values of the parameter A give rise to larger-amplitude and wider dark solitons. Unlike a dark soliton associated with usual cubic Schrödinger equation in which the group dispersion and the nonlinearity coefficient have opposite sign, the modulus of the wave function in the present work has localized maxima on both sides of the density depletion. If the boundary conditions are shifted below 180 degrees (i.e. by a complex number), we have a "grey soliton" which is characterized by a non-zero density at the center of the soliton. In order to assess the dynamics and stability of the dark soliton, we have solved the time-dependent system of Eqs. (1) and (2) numerically, and

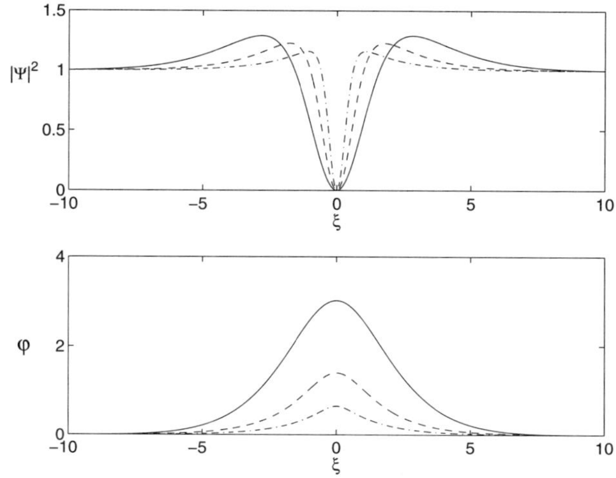

Fig. 1. The electron density $|\Psi|^2$ (the upper panel) and electrostatic potential φ (the lower panel) associated with a dark soliton supported by the system of equations (4) and (5), for $A = 5$ (solid lines), $A = 1$ (dashed lines), and $A = 0.2$ (dash-dotted line). After Ref. 19.

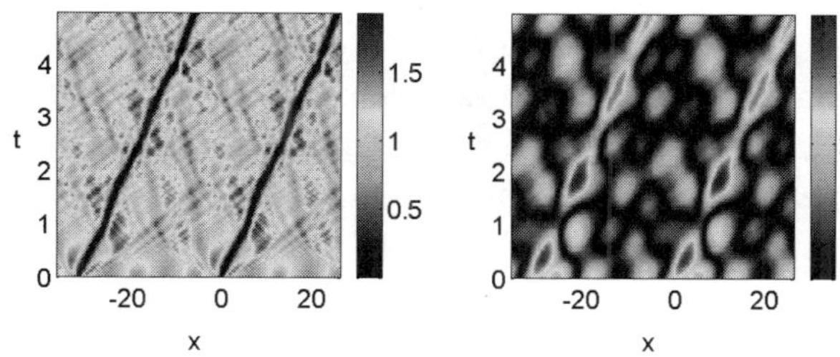

Fig. 2. The time-development of the electron density $|\Psi|^2$ (left-hand panel) and electrostatic potential φ (the right-hand panel), obtained from a simulation of the system of equations (1) and (2). The initial condition is $\Psi = 0.18 + \tanh[20\sin(x/10)]\exp(iKx)$, with $K = v_0/2A$, $A = 5$ and $v_0 = 5$. After Ref. 19.

have displayed the result in Fig. 2. The initial condition is $\Psi = 0.18 + \tanh[20\sin(x/10)]\exp(iKx)$, where $K = v_0/2A$, $A = 5$ and $v_0 = 5$. We clearly see oscillations and wave turbulence in the time-dependent solution presented in Fig. 2. Two very clear and long-lived dark solitons are visible, associated with a positive potential of $\varphi \approx 3$, which is consistent with the quasi-stationary solution of Fig. 1 for $A = 5$. Hence, the dark solitons seem to be robust structures that can withstand perturbations and turbulence during a considerable time.

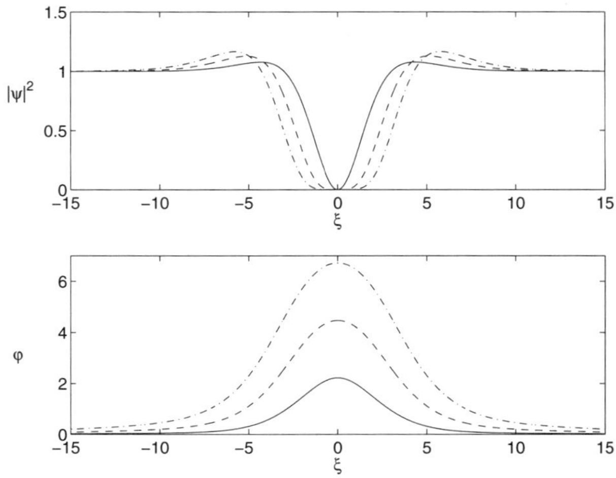

Fig. 3. The electron density $|\Psi|^2$ (upper panel) and electrostatic potential φ (lower panel) associated with a two-dimensional vortex supported by the system (7) and (8), for the charge states $n = 1$ (solid lines), $n = 2$ (dashed lines) and $n = 3$ (dash-dotted lines). We used $A = 5$ in all cases. After Ref. 19.

We next consider two-dimensional vortex structures of the form $\Psi = \psi(r)\exp(in\theta - i\Omega t)$, where r and θ are the polar coordinates defined via $x = r\cos(\theta)$ and $y = r\sin(\theta)$, Ω is a constant frequency shift, and $n = 0, \pm 1, \pm 2, \ldots$ for different excited states (charge states). With this, we can write Eqs. (1) and (2) in the form

$$\Omega\psi + A\left(\frac{d^2}{dr^2} + \frac{1}{r}\frac{d}{dr} - \frac{n^2}{r^2}\right)\psi + \varphi\psi - |\psi|^4\psi = 0, \tag{7}$$

and

$$\left(\frac{d^2}{dr^2} + \frac{1}{r}\frac{d}{dr}\right)\varphi = |\psi|^2 - 1, \tag{8}$$

respectively, where the boundary conditions $\psi = 1$ and $\varphi = d\psi/dr = 0$ at $r = \infty$ determine $\Omega = 1$. Different signs of n describe different rotation directions of the vortex. For $n \neq 0$, we must have $\psi = 0$ at $r = 0$, and from symmetry considerations we have $d\varphi/dr = 0$ at $r = 0$. In Fig. 3, we display numerical solutions of Eqs. (7) and (8) for different charge states n and for $A = 5$. We see that the vortex is characterized by a complete depletion of the electron density at the core of the vortex, and is associated with a positive electrostatic potential. In order to assess the stability of the vortices, we have numerically solved the time-dependent system of Eqs. (1)

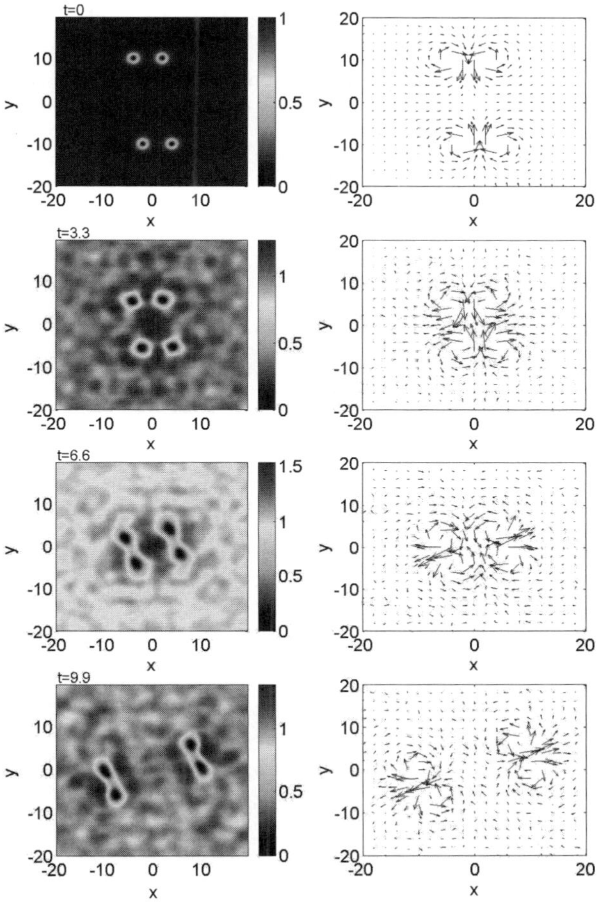

Fig. 4. The electron density $|\Psi|^2$ (left panel) and an arrow plot of the electron current $i\,(\Psi\nabla\Psi^* - \Psi^*\nabla\Psi)$ (right panel) associated with singly charged ($n = 1$) two-dimensional vortices, obtained from a simulation of the time-dependent system of equations (1) and (2), at times $t = 0$, $t = 3.3$, $t = 6.6$ and $t = 9.9$ (upper to lower panels). We used $A = 5$ The singly charged vortices form pairs and keep their .dentities. After Ref. 19.

and (2) in two-space dimensions for singly charged vortices and presented our results in Fig. 4. We have placed four vortex-like structures at some distance from each other, by the initial condition $\Psi = f_1 f_2 f_3 f_4$, where $f_j = \tanh[\sqrt{(x-x_j)^2 + (y-y_j)^2}]\exp[+in\arg(x-x_j, y-y_j)]$. Here $(x_1, y_1) = (-4, 10)$, $(x_2, y_2) = (2, 10)$, $(x_3, y_3) = (-2, -10)$, and $(x_4, y_4) = (4, -10)$. The function $\arg(x, y)$ denotes the angle between the x axis and the point (x, y), and it takes values between $-\pi$ and π. The initial conditions are such

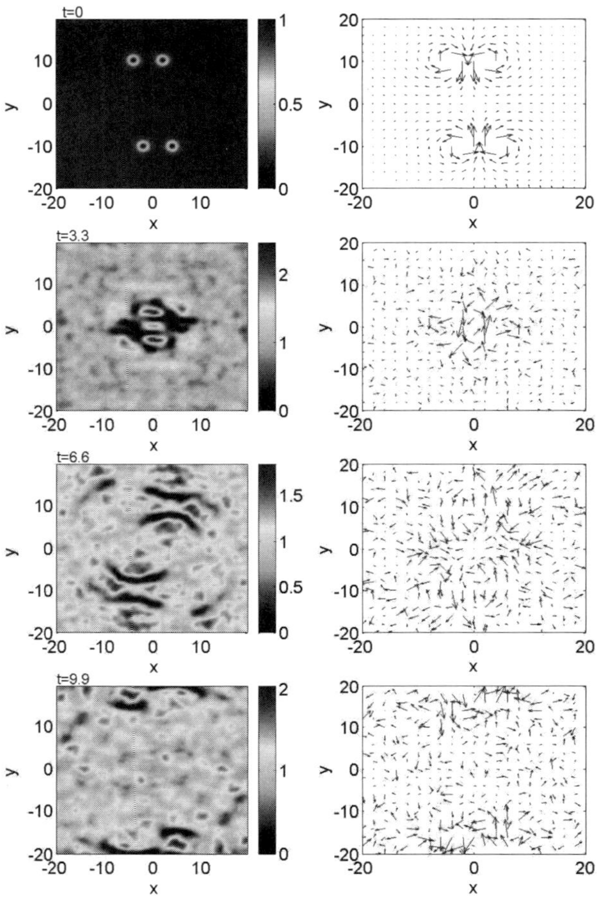

Fig. 5. The electron density $|\Psi|^2$ (left panel) and an arrow plot of the electron current $i\,(\Psi\nabla\Psi^* - \Psi^*\nabla\Psi)$ (right panel) associated with double charged ($n = 2$) two-dimensional vortices, obtained from a simulation of the time-dependent system of Eqs. (1) and (2), at times $t = 0$, $t = 3.3$, $t = 6.6$ and $t = 9.9$ (upper to lower panels). We used $A = 5$. The doubly charged vortices dissolve into nonlinear structures and wave turbulence. After Ref. 19.

that the vortices are organized in two vortex pairs, as seen in the upper panels of Fig. 4. The vortices in the pairs have opposite polarity on the rotation, as seen in the electron fluid rotation direction in the upper right panel. The time-development of the system exhibits that the "partners" in the vortex pairs attract each other and propagate together with a constant velocity. When the two vortex pairs collide and interact (see the second and third pairs of panels in Fig. 4), the vortices keep their identities and change partners in a manner of asymptotic freedom, resulting into two new vortex pairs which propagate obliquely to the original propagation direction. For vortices that are multiply charged ($|n| > 1$), we have a breakup of the vortices and the formation of quasi one-dimensional dark solitons and pairs of vortices with single charge states. One such example is shown in Fig. 5, where we have simulated the system of Eqs. (1) and (2), with the same initial condition as the one in Fig. 4, except that we here have taken $n = 2$ to make the vortices doubly charged. The second row of panels in Fig. 5 reveals that the vortex pairs keep their identities for some time, while a quasi one-dimensional density cavity is formed between the two vortex pairs. At a later stage, the four vortices dissolve into complicated nonlinear structures and wave turbulence. Hence, the nonlinear dynamics is very different between singly and multiply charged solitons, where only singly charged vortices are long-lived and keep their identities. This is in line with previous results on the nonlinear Schrödinger equation, where it was noted that vortices with higher charge states are unstable [17]. In the numerical simulations of Eqs. (1) and (2), we used a pseudo-spectral method to approximate the x and y derivatives and a fourth-order Runge-Kutta scheme for the time-stepping. The numerical simulations confirmed the conservation laws of the electron number, momentum and energy up to the accuracy of the numerical scheme. The numerical solutions of the time-independent systems (4)–(5) and (7)–(8) were obtained by using the Newton method, where the ξ derivatives were approximated with a second-order centered difference scheme with appropriate boundary conditions on Ψ and φ.

3. Interaction between intense electromagnetic waves and quantum plasma oscillations

In this section, we discuss the nonlinear interaction between intense electromagnetic radiation and quantum plasma oscillations [35]. We consider a one-dimensional geometry of an unmagnetized dense electron-ion plasma, in which immobile ions form the neutralizing background. Thus, we are investigating the phenomena on a timescale shorter than the ion plasma period. Our dense quantum plasma contains an intense circularly polarized electromagnetic (CPEM) plane wave that nonlinearly interacts with EPOs. The nonlinear interaction between intense CPEM waves and EPOs gives rise to an envelope of the CPEM vector potential $\mathbf{A}_\perp = A_\perp(\hat{\mathbf{x}} + i\hat{\mathbf{y}})\exp(-i\omega_0 t + ik_0 z)$, which obeys the nonlinear Schrödinger equation [34]

$$2i\Omega_0\left(\frac{\partial}{\partial t} + V_g\frac{\partial}{\partial z}\right)A_\perp + \frac{\partial^2 A_\perp}{\partial z^2} - \left(\frac{|\psi|^2}{\gamma} - 1\right)A_\perp = 0, \tag{9}$$

where the electron wave function ψ and the scalar potential are governed by, respectively,

$$iH\frac{\partial\psi}{\partial t} + \frac{H^2}{2}\frac{\partial^2\psi}{\partial z^2} + (\phi - \gamma + 1)\psi = 0, \tag{10}$$

and

$$\frac{\partial^2\phi}{\partial z^2} = |\psi|^2 - 1, \tag{11}$$

where $\Omega_0 = \omega_0/\omega_{pe}$, $V_g = v_g/c$, $H = \hbar\omega_{pe}/mc^2$, $v_g = k_0c^2/\omega_0$ is the group velocity of the CPEM waves, and $\gamma = (1 + |A_\perp|^2)^{1/2}$ is the relativistic gamma factor due to the electron quiver velocity in the CPEM wave fields. Furthermore, $\omega_0 = (k_0^2c^2 + \omega_{pe}^2)^{1/2}$ is the CPEM wave frequency, k_0 is the wavenumber, c is the speed of light in vacuum, $\omega_{pe} = (4\pi n_0 e^2/m)^{1/2}$ is the electron plasma frequency, e is the magnitude of the electron charge, n_0 is the equilibrium electron number density, and m is the electron rest mass. In (9)–(11) the time and space variables are normalized by the inverse electron plasma frequency ω_{pe}^{-1} and skin depth $\lambda_e = c/\omega_{pe}$, respectively, the scalar potential ϕ by mc^2/e, the vector potential $A_\perp$ by mc^2/e, and the electron wave function $\psi(z,t)$ by $n_0^{1/2}$. The nonlinear coupling between intense CPEM waves and EPOs comes about due to the nonlinear current density, which is represented by the term $|\psi|^2A_\perp/\gamma$ in Eq. (9). The electron number density is defined as $n_e = \psi\psi^* = |\psi|^2$, where the asterisk denotes the complex conjugate. In Eq. (10), $1-\gamma$ is the relativistic ponderomotive potential [34], which arises due to the cross-coupling between the CPEM wave-induced electron quiver velocity and the CPEM wave magnetic field. The second term in the left-hand side in (10) is associated with the quantum Bohm potential [12].

It is well known [36] that a relativistically strong electromagnetic wave in a classical electron plasma is subjected to the Raman scattering and modulational instabilities. At quantum scales, these instabilities will be modified by the dispersive effects caused by the tunnelling of the electrons. In order to investigate the quantum mechanical effects on the relativistic parametric instabilities in a dense quantum plasma in the presence of a relativistically strong CPEM pump wave, we let $\phi(z,t) = \phi_1(z,t)$, $A_\perp(z,t) = [A_0 + A_1(z,t)]\exp(-i\alpha_0 t)$ and $\psi(z,t) = [1+\psi_1(z,t)]\exp(-i\beta_0 t)$, where A_0 is the large-amplitude CPEM pump and A_1 is the small-amplitude fluctuations of the CPEM wave amplitude due to the nonlinear coupling between CPEM waves and EPOs, i.e. $|A_1| \ll |A_0|$, and ψ_1 ($\ll 1$) is the small-amplitude perturbations in the electron wave function. The constant frequency shifts, determined from Eqs. (9) and (10), are $\alpha_0 = (1/\gamma_0 - 1)/(2\Omega_0)$ and $\beta_0 = (1-\gamma_0)/H$, where $\gamma_0 = (1+|A_0|^2)^{1/2}$. The first-order perturbations in the electromagnetic vector potential and the electron wave function are expanded into their respective sidebands as $A_1(z,t) = A_+ \exp(iKz - i\Omega t) + A_- \exp(-iKz + i\Omega t)$ and $\psi_1(z,t) = \psi_+ \exp(iKz - i\Omega t) + \psi_- \exp(-iKz + i\Omega t)$, while the potential is expanded as $\phi(z,t) = \widehat{\phi}\exp(iKz - i\Omega t) + \widehat{\phi}^* \exp(-iKz + i\Omega t)$, where Ω and K are the frequency and wave number of the electron plasma oscillations, respectively. Inserting the above mentioned Fourier ansatz into Eqs. (9)–(11), linearizing the resultant system of equations, and sorting into equations for different Fourier modes, we obtain the nonlinear dispersion relation

$$1 - \left(\frac{1}{D_+} + \frac{1}{D_-}\right)\left(1 + \frac{K^2}{D_L}\right)\frac{|A_0|^2}{2\gamma_0^3} = 0, \tag{12}$$

where $D_\pm = \mp 2\Omega_0(\Omega - V_g K) + K^2$ and $D_L = 1 + H^2K^4/4 - \Omega^2$. We note that $D_L = 0$ yields the linear dispersion relation $\Omega^2 = 1 + H^2K^4/4$ for the EPOs in a dense quantum plasma [4]. For $H \to 0$ we recover from (12) the nonlinear dispersion relation for relativistically large amplitude electromagnetic waves in a classical electron plasma [36]. The dispersion relation (12) governs the Raman backward and forward scattering instabilities, as well as the modulational instability. In the long wavelength limit $V_g \ll 1$, $\Omega_0 \approx 1$ we introduce the ansatz $\Omega = i\Gamma$, where the normalized (by ω_{pe}) growth rate $\Gamma \ll 1$, and obtain from Eq. (12) the growth rate $\Gamma = (1/2)|K|\{(|A_0|^2/\gamma_0^3)[1 + K^2/(1 + H^2K^4/4)] - K^2\}^{1/2}$ of the modulational instability. For $|K| < 1$ and $H < 1$, the linear growth rate is only weakly depending on the quantum parameter H. However, possible nonlinear saturation of the modulational instability may lead to localized CPEM wave packets, which are trapped in a quantum electron hole. Such localized electromagnetic wavepackets would have length scales much shorter than those involved in the modulational instability process. Here quantum diffraction effects (associated with the quantum Bohm potential) become very important. In order to investigate the quantum diffraction effect on such localized electromagnetic pulses, we consider a steady state structure moving with a constant speed V_g. Inserting the ansatz $A_\perp = W(\xi)\exp(-i\Omega t)$, $\psi = P(\xi)\exp(ikx - i\omega t)$ and $\phi = \phi(\xi)$ into Eqs. (9)–(11), where $\xi = z - V_g t$, $k = V_g/H$ and $\omega = V_g^2/2H$, and where $W(\xi)$ and $P(\xi)$ are real, we obtain from (9)-(11) the coupled system of equations

$$\frac{\partial^2 W}{\partial \xi^2} + \left(\lambda - \frac{P^2}{\gamma} + 1\right) W = 0, \tag{13}$$

$$\frac{H^2}{2}\frac{\partial^2 P}{\partial \xi^2} + (\phi - \gamma + 1)P = 0, \tag{14}$$

where $\gamma = (1 + W^2)^{1/2}$, and

$$\frac{\partial^2 \phi}{\partial \xi^2} = P^2 - 1, \tag{15}$$

with the boundary conditions $W = \Phi = 0$ and $P^2 = 1$ at $|\xi| = \infty$. In Eq. (13), $\lambda = 2\Omega_0\Omega$ represents a nonlinear frequency shift of the CPEM wave. In the limit $H \rightarrow 0$, we have from (14) $\phi = \gamma - 1$, where $P \neq 0$, and we recover the classical (non-quantum) case of the relativistic solitary waves in a cold plasma [37]. We note that the system of equations (13)–(15) admits a Hamiltonian

$$\begin{aligned} Q_H =& \frac{1}{2}\left(\frac{\partial W}{\partial \xi}\right)^2 + \frac{H^2}{2}\left(\frac{\partial P}{\partial \xi}\right)^2 - \frac{1}{2}\left(\frac{\partial \phi}{\partial \xi}\right)^2 \\ &+ \frac{1}{2}(\lambda + 1)W^2 + P^2 - \gamma P^2 + \phi P^2 - \phi = 0, \end{aligned} \tag{16}$$

where we have used the boundary conditions $\partial/\partial\xi = 0$, $W = \phi = 0$ and $|P| = 1$ at $|\xi| = \infty$.

In order to asses the importance of our investigation, we now present numerical solutions of (8)–(11) and (13)–(15), ensuring that (16) is conserved. We chose parameters that are representative of the next generation laser-based plasma compression (LBPC) schemes [24,?]. The formula [34] $eA_\perp/mc^2 = 6 \times 10^{-10}\lambda_s\sqrt{I}$ will determine the normalized vector potential, provided that the CPEM wavelength λ_s (in microns) and intensity I (in W/cm^2) are known. It is expected that in LBPC schemes, the electron number density n_0 may reach 10^{27} cm^{-3} and beyond, and the peak values of $eA_\perp/mc^2$ may be in the range 1-2 (e.g. for focused EM pulses with $\lambda_s \sim 0.15$ nm and $I \sim 5 \times 10^{27}$ W/cm^2). For $\omega_{pe} = 1.76 \times 10^{18}$ s^{-1}, we have $\hbar\omega_{pe} = 1.76 \times 10^{-9}$ erg and $H = 0.002$, since $mc^2 = 8.1 \times 10^{-7}$ erg. The electron skin depth $\lambda_e \sim 1.7$ Å. On the other hand, a higher value of $H = 0.007$ is achieved for $\omega_{pe} = 5.64 \times 10^{18}$ s^{-1}. Thus, our numerical solutions below, based on these two values of H, have focused on scenarios that are relevant for the next generation intense laser-solid density plasma interaction experiments [24].

We first numerically solved Eqs. (13)–(15) for several values of H. Here, we solved the nonlinear boundary value problem with the boundary conditions $W = \phi = 0$ and $P = 1$ at the boundaries at $\xi = \pm 10$. We used centered second-order approximations for the second derivatives and solved the obtained nonlinear system of equations numerically by using the Newton method. The results are displayed in Figs. 6 and 7. We see that the solitary envelope pulse is composed of a single maximum of the localized vector potential W and a local depletion of the electron density P^2, and a localized positive potential ϕ at the center of the solitary pulse. The latter has a continuous spectrum in λ, where larger values of negative λ are associated with larger amplitude solitary EM pulses. At the center of the solitary EM pulse, the electron density is partially depleted, as in panels a) of Fig. 6, and for larger amplitudes of the EM waves we have stronger depletion of the electron density, as shown in panels b) and c) of Fig. 6. For cases where the electron density goes to almost zero in the classical case [37], one important quantum effect is that the electrons can tunnel into the depleted region. This is seen in Fig. 7, where the electron density remains nonzero for the larger value of H in panels a), while the density shrinks to zero for the smaller value of H in panel b).

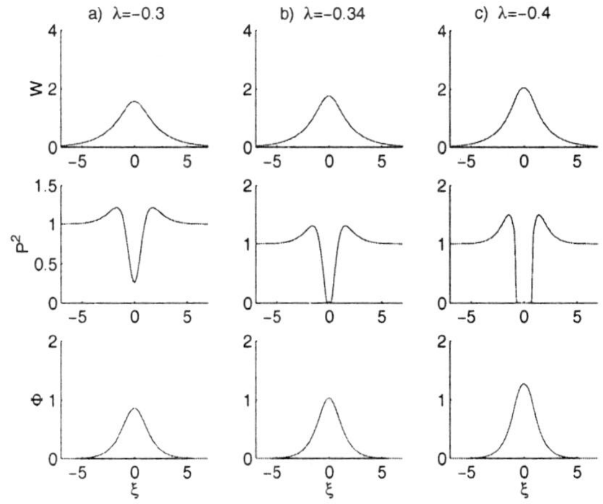

Fig. 6. The profiles of the CPEM vector potential $A_\perp$, the electron number density and the scalar potential (upper to lower rows of panels) for $\lambda = -0.3$, $\lambda = -3.4$ and $\lambda = -0.4$, with $H = 0.002$. After Ref. 35.

In order to investigate the quantum diffraction effects on the dynamics of localized CPEM wavepackets, we have solved the system of Eqs. (9)–(11) numerically. We considered the long-wavelength limit $\omega_0 \approx 1$ and $V_g \approx 0$. In the initial conditions, we use an EM pump with a constant amplitude $A_\perp = A_0 = 1$ and a uniform plasma density $\psi = 1$. A small amplitude noise (random numbers) of order 10^{-2} is added to $A_\perp$ to give a seed for

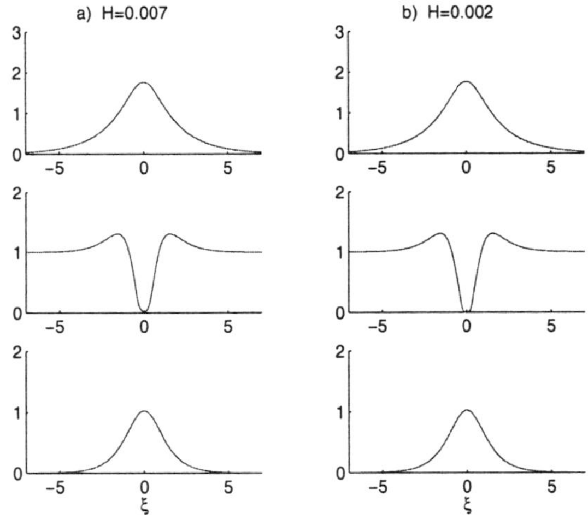

Fig. 7. The profiles of the CPEM vector potential $A_\perp$, the electron number density and the scalar potential (upper to lower rows of panels) for $H = 0.007$ and $H = 0.002$, with $\lambda = -0.34$. After Ref. 35.

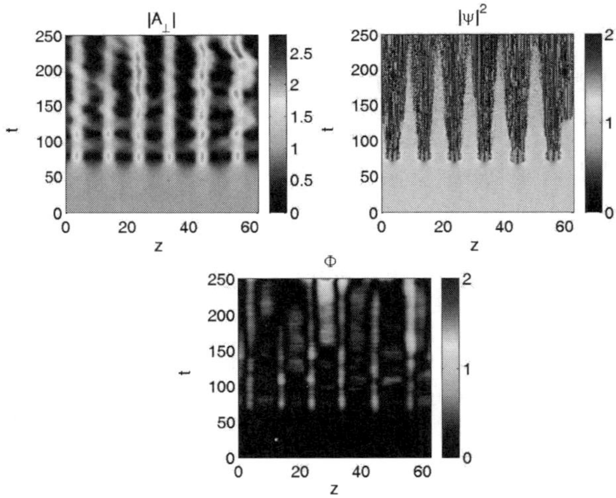

Fig. 8. The dynamics of the CPEM vector potential $A_\perp$ and the electron number density $|\psi|^2$ (upper panels) and of the electrostatic potential Φ (lower panel) for $H = 0.002$. After Ref. 35.

any instability. The numerical results are displayed in Figs. 8 and 4 for $H = 0.002$ and $H = 0.007$, respectively. In both cases, we see an initial

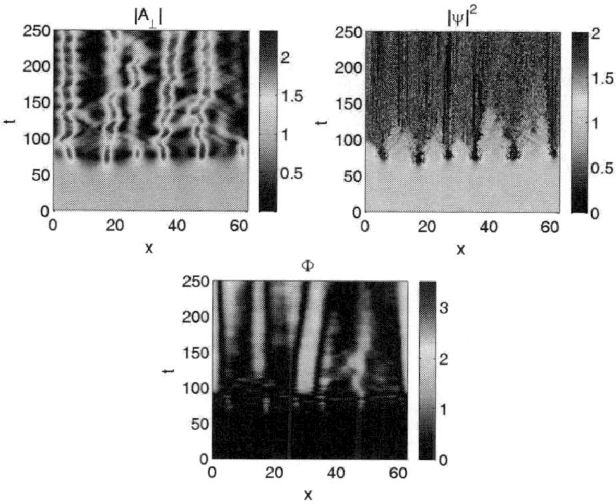

Fig. 9. The dynamics of the CPEM vector potential $A_\perp$ and the electron number density $|\psi|^2$ (upper panels) and the electrostatic potential ϕ (lower panel) for $H = 0.007$. After Ref. 35.

linear growth phase and a wave collapse at $t \approx 70$, in which almost all the CPEM wave energy is contracted into a few well separated localized CPEM wave pipes. These are characterized by a large bell-shaped amplitude of the CPEM wave, an almost complete depletion of the electron number density at the center of the CPEM wavepacket, and a large-amplitude positive electrostatic potential. Comparing Fig. 8 with Fig. 9, we see that there is a more complex dynamics in the interaction between the CPEM wavepackets for the larger $H = 0.007$, shown in Fig. 9, in comparison with $H = 0.002$, shown in Fig. 8, where the wavepackets are almost stationary when they are fully developed. We have here neglected the effects of the ion dynamics. The latter may be important for the development of expanding plasma bubbles (cavities) on longer timescales (e.g. the ion plasma period) [38].

4. Conclusions

In summary, we have demonstrated the existence of localized nonlinear structures in quantum electron plasmas. The electron dynamics in the latter is governed by a coupled nonlinear Schrödinger and Poisson system of equations, which admit a set of conserved quantities (the total number of electrons, the electron momentum, the electron angular momentum, and the electron energy). The latter were checked numerically. Quasi-stationary, localized structures in the form of one-dimensional dark solitons and two-dimensional vortices were found by solving the time-independent coupled system of equations numerically. These structures are associated with a local depletion of the electron density associated with positive electrostatic potential, and are parameterised by the quantum coupling parameter only. In the two-dimensional geometry, we have a class of vortices of different excited states (charge states) associated with a complete depletion of the electron density and an associated positive potential. The numerical simulation of the time-dependent system of equations shows the formation of stable dark solitons in one-space dimension with an amplitude consistent with the one found from the time-independent solutions. In two-space dimensions, the dark solitons of the first excited state were found to be stable and the preferred nonlinear state was in the form of vortex pairs of vortices with different polarities. One-dimensional dark solitons and singly charge two-dimensional vortices are thus long-lived nonlinear structures, which can transport information at quantum scales in micro-mechanical systems and dense laboratory plasmas. We have also presented theoretical and computer simulation studies of nonlinearly interacting intense CPEM waves and EPOs in very dense quantum plasmas. The localized dark solitons, vortices, and CPEM wave structures, as discussed here, may be useful for information transfer as well as for electron acceleration in dense quantum plasmas.

References

1. P. A. Markowich *et al.*, *Semiconductor Equations* (Springer, Berlin, 1990).
2. K. H. Becker, K. H. Schoenbach, and J. G. Eden, *J. Phys. D: Appl. Phys.* **39**, R55 (2006).
3. M. Opher *et al.*, *Phys. Plasmas* **8**, 2454 (2000).
4. D. Pines, *J. Nucl. Energy: Part C: Plasma Phys.* **2**, 5 (1961).
5. F. Haas, G. Manfredi, and M. Feix, *Phys. Rev. E* **62**, 2763 (2000).
6. G. Manfredi and F. Haas, *Phys. Rev. B* **64**, 075316 (2001).
7. D. Anderson *et al.*, *Phys. Rev. E* **65**, 046417 (2002).
8. F. Haas, L. G. Garcia, J. Goedert, and G. Manfredi, *Phys. Plasmas* **10**, 3858 (2003)
9. F. Haas, *Phys. Plasmas* **12**, 062117 (2005).
10. A. V. Andreev, *JETP Lett.* **72**, 238 (2000).
11. G. Manfredi, arXiv:quant-ph/0505004-v1 (2005).
12. C. L. Gardner and C. Ringhofer, *Phys. Rev. E* **53**, 157 (1996).
13. M. Loffredo and L. Morato, *Nuovo Cimento Soc. Ital Fis. B* **108B**, 205 (1993).
14. R. Feynman, *Statistical Mechanics, A Set of of Lectures* (Benjamin, Reading, 1972).

15. A. Domps *et al.*, *Phys. Rev. Lett.* **80**, 5520 (1998).
16. E. B. Kolomeisky *et al.*, *Phys. Rev. Lett.* **85**, 1146 (2000).
17. I. A. Ivonin, V. P. Pavlenko, and H. Persson, *Phys. Rev. E* **60**, 492 (1999).
18. G. Manfredi, *Fields Inst. Commun.* **46**, 263 (2005).
19. P. K. Shukla and B. Eliasson, *Phys. Rev. Lett.* **96** 245001 (2006).
20. M. Marklund and G. Brodin, *Phys. Rev. Lett.* **98**, 025001 (2007).
21. S. X. Hu and C. H. Keitel, *Phys. Rev. Lett.* **83**, 4709 (1999).
22. Y. A. Salamin *et al.*, *Phys. Rep.* **427**, 41 (2006).
23. S. H. Glenzer *et al.*, *Phys. Rev. Lett.* **98**, 065002 (2007).
24. V. M. Malkin *et al.*, *Phys. Rev. E* **75**, 026404 (2007).
25. H. Azechi *et al.*, *Plasma Phys. Control. Fusion* **48**, B267 (2006).
26. O. G. Benvenuto and M. A. De Vito, *Mon. Not. R. Astron. Soc.* **362**, 891 (2005).
27. G. Chabrier *et al.*, *J. Phys.: Condens. Matter* **14**, 9133 (2002).
28. G. Chabrier *et al.*, *J. Phys. A: Math. Gen.* **39**, 4411 (2006).
29. Y.Y. Lau *et al.*, *Phys. Rev. Lett.* **66**, 1446 (1991).
30. L. K. Ang *et al.*, *Phys. Rev. Lett.* **91**, 208303 (2003).
31. L. K. Ang and P. Zhang, *Phys. Rev. Lett.* **98**, 164802 (2007).
32. G. Mourou *et al.*, *Rev. Mod. Phys.* **78**, 309 (2006).
33. M. Marklund and P. K. Shukla, *Rev. Mod. Phys.* **78**, 591 (2006).
34. P.K. Shukla *et al.*, *Phys. Rep.* **138**, 1 (1986).
35. P. K. Shukla and B. Eliasson, *Nonlinear interactions between electromagnetic waves and electron plasmas oscillations in quantum plasmas*, *Phys. Rev. Lett.* **99**, 06xxxx (in press, 2007).
36. C. J. McKinstrie and R. Bingham, *Phys. Fluids B* **4**, 2626 (1992).
37. J. H. Marburger and R. F. Tooper, *Phys. Rev. Lett.* **35**, 1001 (1975).
38. M. Borghesi *et al.*, *Phys. Rev. Lett.* **88**, 135002 (2002).

Vortex in plasmas – topology, singularity and scale hierarchy

Z. Yoshida

Graduate School of Frontier Sciences, The University of Tokyo, Chiba 277-8561, Japan

The Lagrangian for collective (fluid-mechanical) matter-field coupling must incorporate a non-canonical structure that is symbolized by the nonzero "vorticity" of the momentum (violating the basic Hamilton-Jacobi relation $\boldsymbol{P} = \nabla S$ for the momentum $\boldsymbol{P}$ and the action S). The Eulerian view of the mechanics, describing the dynamics in terms of "synchronic" variables, reveals the topological defects of the Poisson bracket, which are represented by the so-called "Casimir invariants". Adding Casimir invariants, one may transform the Hamiltonian without changing the dynamics. The critical points of the transformed Hamiltonian describes multi-scale matter-field couplings.

Keywords: Vortex, Nonlinear field theory, Topological defect, Scale hierarchy.

1. Introduction

From the view-point of nonlinear field theory, "flow" is a delicate subject —the nature of flow is often indicated by adjectives crowned with "non", such as non-linear, non-Hermitian, non-canonical, non-Abelian, non-commutative, etc., and thus, fluid mechanics is a rich resource of basic concepts that may be disseminated to other fields of sciences.

The origin of "vorticity" is the central problem which makes the theory extremely non-trivial. The Lagrangian for fluid motion must incorporate a "non-canonical" structure that is symbolized by the vorticity of the momentum. Remembering the basic relation $\boldsymbol{P} = \nabla S$ ($\boldsymbol{P}$ is the momentum and S is the action) of the Hamilton-Jacobi equations, it is interesting to ask how a mechanical system, being described by the criticality of some action, can attain a vorticity ($\nabla \times \boldsymbol{P} \neq 0$). A potential flow $\boldsymbol{P} = \nabla S$ suffices for the mechanics of a "particle", because the notion of a bundle of orbits is not there; the vorticity (helicity) is a geometric index measuring the twists of multiple orbits. It is worthwhile noting that any theory that constructs fields from the action cannot have a vorticity of the phase; a well-known example is the Ginzburg-Landau wave function of a super-fluid.

The attempt to formulate a fluid-mechanical Lagrangian has a long history.[1–4] Some different Lagrangians have been proposed, while some of them are "incomplete" to describe general motion of fluids, and the relations among these formulations are not well understood. It is rather straightforward to extend the classical-mechanical Lagrangian for particle motion to that for continuum by invoking the notion of diffeomorphism that describes the displacements of matter in the "Lagrangian view". However, difficult exploration has been continuing aiming at developing an "Eulerian view". While classical mechanics is based on the Lagrangian view, standard field theories stand behind the Eulerian view. The above-mentioned problem, the origin of vorticity is, indeed, posed in the quest for the latter.

The relation between the Lagrangian and Eulerian views may be compared with the "diachrony" and "synchrony" in the structuralistic/post-structuralistic terminology. The Lagrangian view attributes the origin of vorticity to the "initial condition", and its diachronic analysis of vortex dynamics is unaware of the "topological defects" associated with the vorticity. The Eulerian view finds singularities in formulating Poisson brackets with synchronic variables, revealing the deep structure of the nonlinear dynamics. So-called "Casimir invariants",[5,6] defining the kernel of the Poisson bracket operator, produces a variety of interesting structures. We will formulate a variational principle to analyze multi-scale collective matter-field coupling in plasmas.[7,8]

2. Fluid Mechanics in Lagrangian View

The Lagrangian for the collective matter-field couplings may be naturally derived by generalizing the Lagrangian of a particle (mass m and charge e) in the presence of an EM field (we consider the non-relativistic limit for simplicity):

$$L = L_P + L_{EM}, \tag{1}$$

$$L_P = \boldsymbol{P} \cdot \boldsymbol{v} - H, \tag{2}$$

$$L_{EM} = \int \mathcal{L}_{EM} dx = \int -\frac{1}{4} F_{\mu\nu} F^{\mu\nu} dx, \tag{3}$$

where $H = p^2/(2m) + e\phi$ is the Hamiltonian, $\boldsymbol{P} = \boldsymbol{p} + (e/c)\boldsymbol{A}$ is the canonical momentum, and $\boldsymbol{p}$ is the mechanical momentum. The vector and the scalar potentials define the four potential $A^\mu = (\phi, \boldsymbol{A})$ whose curl is the Faraday (field strength) tensor $F_{\mu\nu} = \partial_\mu A_\nu - \partial_\nu A_\mu$. The velocity $\boldsymbol{v}$ is related to the particle orbit $\boldsymbol{q}(t)$ (along which L_P is to be evaluated) by

$$\boldsymbol{v} = \dot{\boldsymbol{q}}. \tag{4}$$

The relation (4) is the essential input that "causes" the motion of the particle. Indeed, if we calculate the variation of $\int L\,dt$ for general $\delta \boldsymbol{v}$, we find $\boldsymbol{P} = 0$, but, for $\delta \dot{\boldsymbol{q}}$, with fixed end points $\boldsymbol{q}(t_0)$ and $\boldsymbol{q}(t_1)$, the well-known Lagrange's equation of motion follows.

Generalizing the single particle orbit $\boldsymbol{q}(t)$ to a diffeomorphism $\boldsymbol{Q}(\boldsymbol{x}_0, t)$ in $\mathbf{R}^3$ ($\boldsymbol{x}_0$ being the initial position of each streamline), we construct the fluid Lagrangian

$$L_F = \int \mathcal{L}_F dx = \int (\boldsymbol{P} \cdot \boldsymbol{V} - H_F)\,\rho dx \tag{5}$$

by replacing the single-particle velocity $\dot{\boldsymbol{q}}$ in (4) by the flow velocity

$$\boldsymbol{V}(\boldsymbol{x}, t) = \dot{\boldsymbol{Q}}|_{\boldsymbol{x},t} = \frac{d}{dt}\boldsymbol{Q}(\boldsymbol{x}_0(\boldsymbol{x}, t), t). \tag{6}$$

Here, $\boldsymbol{x}_0(\boldsymbol{x}, t)$ is the initial position of the streamline being connected to the time-space position $(\boldsymbol{x}, t)$. To relate $(\boldsymbol{x}, t)$ to $(\boldsymbol{x}_0, 0)$, one needs the inverse map $\boldsymbol{x}_0(\boldsymbol{x}, t) \equiv \boldsymbol{Q}^{-1}(\boldsymbol{x}, t)$ of the diffeomorphism $\boldsymbol{Q}(\boldsymbol{x}_0, t)$, which traces-back the streamlines (the relations among $\boldsymbol{x} = \boldsymbol{Q}(\boldsymbol{x}_0, t)$, $\boldsymbol{x}_0 = \boldsymbol{Q}^{-1}(\boldsymbol{x}, t)$, and $\boldsymbol{V}$ will be discussed more explicitly in sec. 4).

The fluid Hamiltonian H_F consists of the kinetic and potential energies plus an "internal (thermal) energy" $\varepsilon(\rho)$:

$$H_F = H + \varepsilon(\rho). \tag{7}$$

The fluid density ρ is defined by

$$\rho(\boldsymbol{x}, t) = \rho_0(\boldsymbol{x}_0(\boldsymbol{x}, t)) \cdot \frac{D(\boldsymbol{x}_0)}{D(\boldsymbol{x})}. \tag{8}$$

When we formally take $\rho_0(\boldsymbol{x}_0) = \delta(\boldsymbol{q}_0 - \boldsymbol{x}_0)$, we recover (1) with $\rho dx dt$ giving the integral along the orbit $\boldsymbol{q}(t)$.

In what follows we denote

$$D_t f = \partial_t f + \boldsymbol{V} \cdot \nabla f, \tag{9}$$

$$D_t^* f = \partial_t f + \nabla \cdot (\boldsymbol{V} f). \tag{10}$$

The mass conservation law

$$D_t^* \rho = 0 \tag{11}$$

is a direct consequence of (8). By the criticality of the action $\int (L_F + L_{EM}) dt$, fixing the space-time boundaries, we obtain, from the variation $\delta \boldsymbol{p}$, $\boldsymbol{p} = m\boldsymbol{V}$ $(= m\dot{\boldsymbol{Q}})$, and from $\delta \boldsymbol{Q}$, the equation of motion

$$m D_t \boldsymbol{V} = -\nabla h + e\left(\boldsymbol{E} + \frac{1}{c}\boldsymbol{V} \times \boldsymbol{B}\right), \tag{12}$$

where $\boldsymbol{E} = -\nabla\phi - \partial_t \boldsymbol{A}/c$, $\boldsymbol{B} = \nabla \times \boldsymbol{A}$, and h is the enthalpy density defined by $h = \partial(\rho\varepsilon)/\partial\rho = \varepsilon + \mathcal{P}/\rho$ ($\mathcal{P}$ is the pressure; $\rho\nabla h = \nabla\mathcal{P}$). Maxwell's equations with the currents $(e\rho, e\rho\boldsymbol{V})$ follow from δA_μ.

3. Fluid Mechanics in Eulerian View

In this section, we formulate an Eulerian representation of the fluid Lagrangian. In the Eulerian case where no *a priori* relation between the fluid velocity $\boldsymbol{V}$ and the streamlines is assumed, and the unrestricted variation $\delta\boldsymbol{V}$ yields $\boldsymbol{P} = 0$. To reproduce properly the evolution equations, we must appropriately "constrain" $\boldsymbol{V}$.

The measure $\rho dx dt$, defined by (8), is the generalization of the integral along a single orbit to collective orbits. It was argued, then, that imposing a physically motivated "restriction" on ρ that leads to the conservation law (11), must be a step in the right direction. Serrin, in a pioneering paper,[1] proposed the Lagrangian density

$$\mathcal{L}_F = \rho\,(\boldsymbol{P}\cdot\boldsymbol{V} - H_F) + S D_t^* \rho, \tag{13}$$

in which the variation of the Lagrange multiplier S does exactly that.

The Serrin Lagrangian, however, is inadequate because the momentum $\boldsymbol{P} = \nabla S$ (obtained by the variation $\delta\boldsymbol{V}$) describes only an "irrotational" flow representing a relatively small area of general fluid mechanics. We will see later that this restriction is due to a "canonical structure" premised by (13).

To derive flows with vorticity, Lin[2] imposed another constraint by including a term $\rho\alpha(\partial_t\beta + \boldsymbol{V}\cdot\nabla\beta)$ to the integrand of (13). The resulting flow $\boldsymbol{P} = \nabla S + \alpha\nabla\beta$, corresponding to the so-called Clebsch parameterization of a vector field, does acquire a finite "vorticity" ($\nabla \times \boldsymbol{P} = \nabla\alpha \times \nabla\beta$). Unfortunately, the Clebsch representation (with a single valued S, α and β) is not a global representation for arbitrary flows with non-zero helicity. The Serrin-Lin Lagrangian still falls short of reproducing the fluid dynamics of arbitrary flows.

We propose a generalized and rearranged Serrin-Lin-type Lagrangian density

$$\mathcal{L}_F = \left[\boldsymbol{P}\cdot\boldsymbol{V} - H_F - (D_t S + \lambda^j D_t \sigma_j)\right]\rho. \tag{14}$$

The variational principle $\delta\int(\mathcal{L}_F + \mathcal{L}_{EM})\,dx dt = 0$ yields a complete set of determining equations:

$$\delta \boldsymbol{V} \quad \Rightarrow \quad \boldsymbol{P} = \nabla S + \lambda^j \nabla \sigma_j, \tag{15}$$

$$\delta \boldsymbol{p} \quad \Rightarrow \quad \boldsymbol{p} = m\boldsymbol{V}, \tag{16}$$

$$\delta S \quad \Rightarrow \quad D_t^* \rho = 0, \tag{17}$$

$$\delta \sigma_j \quad \Rightarrow \quad D_t^*(\rho\lambda^j) = 0 \quad \Rightarrow \quad D_t \lambda^j = 0, \tag{18}$$

$$\delta \lambda^j \quad \Rightarrow \quad D_t \sigma_j = 0, \tag{19}$$

$$\delta \rho \quad \Rightarrow \quad D_t S = \boldsymbol{P} \cdot \boldsymbol{V} - (H + h), \tag{20}$$

and, by δA_μ, Maxwell's equations with the currents $(e\rho, e\rho\boldsymbol{V})$. The vector field $\boldsymbol{P} = \nabla S + \lambda^j \nabla \sigma_j$ is now "complete" to represent any three-dimensional vectors if we take (at least) three independent scalars σ_j. The fluid (plasma) equation follows from

$$D_t \boldsymbol{P} = D_t(\nabla S + \lambda^j \nabla \sigma_j) = -\nabla(e\phi + h) + \frac{\epsilon}{c}\left[\boldsymbol{V} \times \boldsymbol{B} + (\boldsymbol{V} \cdot \nabla)\boldsymbol{A}\right],$$

which is equivalent to (12).

The role of S is best understood by referring to the original Serrin form ($\lambda^j = \sigma_j = 0$). Though Serrin's S is a Lagrange multiplier that imposes mass conservation (11), we proffer a different interpretation. By moving (by integrating by parts) D_t^* from ρ to S, one may think of ρ as a Lagrange multiplier demanding that $\mathcal{L}_F$ must be a complete derivative (evaluated through each streamline of $\boldsymbol{V}$) of some scalar field S —this is nothing but Hamilton's principle demanding the criticality of the action integral with S as the "action". Indeed, if the thermal energy ε is neglected and $\lambda^j = \sigma_j = 0$ in (15) and (20), we obtain the well-known Hamilton-Jacobi equations $\partial_t S = -H(\boldsymbol{x}, \boldsymbol{P}, t)$ and $\nabla S = \boldsymbol{P}$.

The additional fields λ^j and σ_j (further constraining the Lagrangian) break the "canonical structure" of the system, yielding "generalized Hamilton-Jacobi equations" (combining (15) and (20)):

$$\partial_t S = -\left[H(\boldsymbol{x}, \boldsymbol{P}, t) + h\right] - \lambda^j \partial_t \sigma_j, \tag{21}$$

$$\nabla S = \boldsymbol{P} - \lambda^j \nabla \sigma_j. \tag{22}$$

We have to specify what we have called "non-canonical". A canonical Hamiltonian mechanics is endowed with a regular (non-degenerate) Poisson bracket, i.e., if $\{F, C\} \equiv 0$ for all F, C must be a constant. Otherwise, i.e., if there exists such a non-constant C, which we call a "Casimir invariant", the system is non-canonical.

The fluid-mechanics equations do have a Casimir invariant, that is the so-called "helicity" defined by $C = \int (\nabla \times \boldsymbol{P}) \cdot \boldsymbol{P} dx$.[5,6] Obviously, if $\boldsymbol{P} =$

∇S, then the helicity density is zero, implying that $C = 0$. A Clebsch-form vector field has a finite helicity density; if $\boldsymbol{P} = \nabla S + \alpha\nabla\beta$, we have $C = \int \nabla\cdot(S\nabla\alpha\times\nabla\beta)dx$. This C is fixed, when we give boundary conditions on the Clebsch parameters S, α, β. Then, the kernel of the Poisson bracket is removed, and hence, we obtain a "canonical" system. Jackiw[4] showed that the (S, ρ) and $(S\alpha, \beta)$ compose canonical pairs describing the fluid motion. As mentioned above, however, Clebsch-form falls short to represent general $\boldsymbol{P}$, and, indeed, the general fluid mechanics is "non-canonical" in a sense that the helicity is a Casimir invariant; See Sec. 5.

4. Connection Between the Lagrangian and Eulerian Views

From (15), we see that the scalar fields λ^j and σ_j may be seen as providing "parameterization" of the vector field $\boldsymbol{P}$ as a 1-form. If the index j runs over the space dimension (= 3), then the vectors $\{\nabla\sigma_j\}$ may no longer be linear independent. Henceforth, we assume both λ^j and σ_j to be the components of three-vectors $\boldsymbol{\lambda}$ and $\boldsymbol{\sigma}$. This interpretation of λ^j and σ_j will now be exploited to relate the Lagrangian and Eulerian formulations of the Lagrangians.

Using the identity $\boldsymbol{V} \equiv D_t\boldsymbol{x}$ (x_j is the Cartesian parameterization), one may write

$$\boldsymbol{P}\cdot\boldsymbol{V} - \boldsymbol{\lambda}\cdot D_t\boldsymbol{\sigma} = \boldsymbol{P}\cdot D_t\boldsymbol{\xi} - \boldsymbol{\mu}\cdot D_t\boldsymbol{\sigma},$$

where $\boldsymbol{\xi} = \boldsymbol{x} - \boldsymbol{\sigma}$ and $\boldsymbol{\mu} = \boldsymbol{\lambda} - \boldsymbol{P}$. Then, (14) transforms to

$$\mathcal{L}_F = (\boldsymbol{P}\cdot D_t\boldsymbol{\xi} - H_F - D_tS - \boldsymbol{\mu}\cdot D_t\boldsymbol{\sigma})\,\rho. \tag{23}$$

From this form, the equivalence between the Eulerian and the previous Lagrangian (5) formalism will be established when we connect $D_t\boldsymbol{\xi}$ with $\dot{\boldsymbol{Q}}$.

In (23), the variation of $\boldsymbol{\mu}$ forces $D_t\boldsymbol{\sigma} = 0$ implying

$$D_t\boldsymbol{\xi} = D_t\boldsymbol{x} \equiv \boldsymbol{V}. \tag{24}$$

This Eulerian representation of the flow velocity is compared with the Lagrangian representation (6) that invoked the diffeomorphism $\boldsymbol{Q}(\boldsymbol{x}_0, t)$. The time derivative $\dot{}$ of $\boldsymbol{Q}(\boldsymbol{x}_0, t)$ is taken for fixed $\boldsymbol{x}_0$ (the initial points of every streamlines). The convective D_t of $\boldsymbol{\xi}$, on the other hand, must be evaluated under the condition $D_t\boldsymbol{\sigma} = 0$. We will now show that such a $\boldsymbol{\sigma}$ (in its simplest choice) is nothing but $\boldsymbol{x}_0$ (hence, $\boldsymbol{\xi} = \boldsymbol{x} - \boldsymbol{\sigma}$ is the "displacement"). The constraint $D_t\boldsymbol{\sigma} = 0$ (i.e., fixing $\boldsymbol{x}_0$), then, allows the identification $\dot{\boldsymbol{Q}} = D_t\boldsymbol{\xi}$.

To show $\boldsymbol{\sigma}(\boldsymbol{x},t) = \boldsymbol{x}_0 = \boldsymbol{Q}^{-1}(\boldsymbol{x},t)$, let us first note that $D_t\boldsymbol{\sigma} = 0$ implies that $\boldsymbol{\sigma}$ is constant along every streamline defined by $\boldsymbol{V}$ and, therefore, must be a function only of the "initial condition" because the initial condition cannot change during the evolution. The simplest expression of such a $\boldsymbol{\sigma}$ is the initial condition itself. Formally, we can construct $\boldsymbol{\sigma}$ by solving $D_t\boldsymbol{\sigma} = 0$ for a given initial condition $\boldsymbol{\sigma}(\boldsymbol{x},0) = \boldsymbol{\sigma}_0(\boldsymbol{x})$. The corresponding "characteristics" equation is

$$\frac{d}{dt}\boldsymbol{x} = \boldsymbol{V}(\boldsymbol{x},t), \quad \boldsymbol{x}(0) = \boldsymbol{x}_0. \tag{25}$$

Solving (25) for every initial value $\boldsymbol{x}_0$, we may construct the diffeomorphism $\boldsymbol{x} = \boldsymbol{Q}(\boldsymbol{x}_0,t)$. Using the inverse map $\boldsymbol{Q}^{-1}(\boldsymbol{x},t)$, we may write $\boldsymbol{\sigma}(\boldsymbol{x},t) = \boldsymbol{\sigma}_0(\boldsymbol{Q}^{-1}(\boldsymbol{x},t))$. The simplest initial condition $\boldsymbol{\sigma}_0(\boldsymbol{x}) = \boldsymbol{x}$, i.e., $\boldsymbol{\sigma}_0$ is the identity, leads to

$$\boldsymbol{\sigma}(\boldsymbol{x},t) = \boldsymbol{Q}^{-1}(\boldsymbol{x},t) \equiv \boldsymbol{x}_0(\boldsymbol{x},t). \tag{26}$$

We already mentioned that when $\boldsymbol{\sigma} = \boldsymbol{x}_0$, $\boldsymbol{\xi} = \boldsymbol{x} - \boldsymbol{\sigma}$ is the Eulerian representation of the "displacement" so that $D_t\boldsymbol{\xi} = \boldsymbol{V} = \dot{\boldsymbol{Q}}$, establishing the equivalence of the Eulerian and Lagrangian descriptions.

5. Casimir Invariants, Beltrami Fields, Lyapunov Function

In this section, we study the non-canonical Hamiltonian structure of the collective matter-field coupling. We start with an abstract formulation. Let $H(u)$ denote a Hamiltonian of a state vector u, and

$$\frac{d}{dt}u = \mathcal{A}\partial_u H(u), \tag{27}$$

where $\mathcal{A}$ is an antisymmetric operator such as $(\mathcal{A}f,g) + (f,\mathcal{A}g) = 0$. [a] Defining the Lie-Poisson bracket by $\{G(u),F(u)\} = (\mathcal{A}\partial_u G(u), \partial_u F(u))$, Liouville's equation is

$$\frac{d}{dt}F(u) = \{H(u),F(u)\}. \tag{28}$$

Simple examples are:

(1) For $u = (q_1,\cdots,q_\nu,p_1,\cdots,p_\nu)$, (27) reads as Hamilton's canonical equation with

$$H(u) = \frac{1}{2m}\sum_j p_j^2 + U(\boldsymbol{q}), \quad \mathcal{A} = \begin{pmatrix} 0 & I \\ -I & 0 \end{pmatrix}.$$

[a] We use the standard notations of the inner product (,) and the norm $\|\ \|$ of the L^2 Hilbert space.

(2) For $u = u(x,t)$, (27) reads as Schrödinger's equation with

$$H(u) = \int \left(\frac{\hbar^2}{2m} |\nabla u|^2 + U(x) \right) dx, \quad \mathcal{A} = -i/\hbar,$$

In what follows, we consider a nonlinear $\mathcal{A}$ that may have a kernel (defining a topological defect). A functional $F(u)$ commuting with $H(u)$ (i.e., $\{H(u), F(u)\} = 0$) is a constant of motion. When $\mathcal{A}$ has a kernel, there may exist a constant of motion originating from the Lie-Poisson bracket (independent to a specific choice of the Hamiltonian), i.e., $C(u)$ such as

$$\{G(u), C(u)\} = 0 \quad (\forall G(u)), \tag{29}$$

which we call a "Casimir invariant". Suppose that the system has Casimir invariants $C_\nu(u)$ ($\nu = 1, \cdots, \nu$). Each Casimir invariant derives a (parameterized) transformation of the Hamiltonian:

$$G_{\boldsymbol{\mu}}(u) = H(u) + \sum_{j=1}^{\nu} \mu_j C_j(u), \tag{30}$$

which does not change the dynamics (28).

While the critical point (equilibrium point) of the Hamiltonian is often trivial, that of the transformed Hamiltonian (30) exhibits interesting structures. Let us analyze the variational principle

$$\delta G_{\boldsymbol{\mu}}(u) = \delta \left(H(u) + \sum_{j=1}^{\nu} \mu_j C_j(u) \right) = 0, \tag{31}$$

which yields a set of critical points parameterized by $\mu_1, \cdots, \mu_\nu$. We call the solutions to (31) as "Beltrami equilibria".[7] Since each $C_j(u)$ may have different scale (order of spacial derivatives), the combination (30) includes multi-scale structures (see Sec. 6).

The stability of a Beltrami equilibrium u_0 is shown by analyzing the Hessian of $G_{\boldsymbol{\mu}}(u)$.[5,9] Writing $u(t) = u_0 + \tilde{u}(t)$, let us derive a conservation law for the perturbation $\tilde{u}(t)$. In the neighbourhood of u_0, we observe

$$\frac{d}{dt}\tilde{u} = \mathcal{L}_0 \mathcal{G}_{\boldsymbol{\mu}} \tilde{u}, \tag{32}$$

where $\mathcal{L}_0$ is the linear operator defined by evaluating $\mathcal{L}$ at $u = u_0$. We note that the contribution from the perturbation $\mathcal{L}'\tilde{u}$ vanishes because of the Beltrami equilibrium condition ($\partial_u G_{\boldsymbol{\mu}}(u_0) = 0$). The operator $\mathcal{G}_{\boldsymbol{\mu}}\tilde{u}$ is the linearization of the $\partial G_{\boldsymbol{\mu}}(u)$, i.e., the Hessian of $G_{\boldsymbol{\mu}}(u)$.

The energy corresponding to the "linearized Hamiltonian" $\mathcal{G}_\mu$ is

$$\tilde{G}_\mu(\tilde{u}) = \frac{1}{2}(\mathcal{G}_\mu u, u), \tag{33}$$

which is a constant of motion governed by (32). This functional $\tilde{G}_\mu(\tilde{u})$ works as a Lyapunov function. A sufficient condition for the stability is the "coercivity":[9]

$$\tilde{G}_\mu(\tilde{u}) \geq c\|\tilde{u}\|^2 \quad (\exists c > 0), \tag{34}$$

which is equivalent to the "convexity" of the transformed Hamiltonian:

$$G_\mu(u_0 + \delta) > G_\mu(u_0) \quad (\forall \delta \neq 0).$$

6. Beltrami Fields –multi-scale flow-field couplings

As mentioned in Sec. 3, the "helicity" of fluid vorticity is a Casimir invariant. In a "plasma" consisting of ion (proton) and electron fluids, both helicities associated with the corresponding flows have distinct length scales, which yield multi-scale Beltrami equilibria.[10]

Here, we assume incompressible condition for simplicity. By subscript $\ell = e$, we identify electrons, and, by $\ell = i$, ions. The (canonical) momentum equations, in an appropriate normalized unit, read as

$$\partial_t \boldsymbol{P}_\ell - \boldsymbol{V}_\ell \times (\nabla \times \boldsymbol{P}_\ell) = -\nabla \Phi_\ell \quad (\ell = e, i), \tag{35}$$

where $\boldsymbol{P}_\ell = m_\ell \boldsymbol{V}_\ell \mp \boldsymbol{A}$ (m_ℓ is the mass) and Φ_ℓ is the energy density of the fluid (sum of the electric, thermal and kinetic energies).

The scale hierarchies in plasmas are highlighted by "singular perturbations".[10] Here, we show how the electron/ion mass ratio yields a singular perturbation; In the electron momentum $\boldsymbol{P}_e$, we may neglect the small mechanical momentum. Then, $\boldsymbol{V}_e$ is no longer a dynamical variable, so we must determine it by another relation. We may write (normalizing the charge density=1) $\boldsymbol{V}_e = \boldsymbol{V}_i - \boldsymbol{J}$ with the current density $\boldsymbol{J}$. Neglecting the displacement current, $\boldsymbol{J} = \nabla \times \boldsymbol{B}$. In this singular limit, we have to take the curl of the electron equation of (35). Then, the Hamiltonian reduces into $H = \left(\|\boldsymbol{P}_i - \boldsymbol{A}\|^2/m_i + \|\boldsymbol{B}\|^2\right)/2$ and the evolution equation reads as

$$\frac{d}{dt}\begin{pmatrix} \boldsymbol{P}_i \\ \boldsymbol{B} \end{pmatrix} = \begin{pmatrix} (\nabla \times \boldsymbol{P}_i) \times \circ & 0 \\ 0 & -\nabla \times [\boldsymbol{B} \times (\nabla \times \circ)] \end{pmatrix} \begin{pmatrix} \partial_{\boldsymbol{P}_i} H \\ \partial_{\boldsymbol{B}} H \end{pmatrix}. \tag{36}$$

Two helicities (Casimir invariants) are $C_e = (\boldsymbol{A}, \boldsymbol{B})$ and $C_i = (\boldsymbol{P}_i, \nabla \times \boldsymbol{P}_i)$

The "Beltrami equilibria" are given by $\delta(H + \mu_1 C_e + \mu_2 C_i) = 0$, whose Euler-Lagrange equations can be cast into the form of

$$(\text{curl} - \lambda_+)(\text{curl} - \lambda_-)\boldsymbol{u} = 0 \quad (\boldsymbol{u} = \boldsymbol{B}, \boldsymbol{V}_i), \tag{37}$$

where $\lambda_\pm$ are constants related to μ_j.[7] Since two operators (curl $- \lambda_\pm$) commute, the general solution of (37) is given by a linear combination of two "Beltrami eigenfunctions",[11] i.e., $\boldsymbol{u} = c_+\boldsymbol{G}_+ + c_-\boldsymbol{G}_-$, where

$$(\text{curl} - \lambda_\pm)\boldsymbol{G}_\pm = 0, \tag{38}$$

and $c_\pm$ are constants (to be determined by the ratios of H, C_e and C_i). Both eigenvalues $\lambda_\pm$ represent the two distinct scales included in the Beltrami equilibria. This model gives the minimum (simplest) representation of the multi-scale flow-field coupling.

Acknowledgements

The author is grateful to Prof. Swadesh M. Mahajan and Dr. Makoto Hirota for useful comments and discussions.

References

1. J. Serrin, Mathematical principles of classical fluid mechanics, in *Encyclopedia of Physics*, (Springer-Verlag, 1959), 125.
2. C.C. Lin, in *Proc. Int. Sch. Phys. "Enrico Fermi" XXI*, (Academic Press, 1963), 93.
3. B.F. Schutz and R. Sorkin, *Ann. Phys.* **107**, 1 (1977).
4. R. Jackiw, *Lectures on Fluid Dynamics –a particle theorist's view of supersymmetric, non-Abelian, noncommutative fluid mechanics and d-branes* (Springer, 2002).
5. V. Arnold and B. Khesin, *Topological Methods in Hydrodynamics* (Springer-Verlag, Berlin, 1998).
6. P.J. Morrison and J.M. Greene, *Phys. Rev. Lett.* **45**, 790 (1980).
7. S. M. Mahajan and Z. Yoshida, *Phys. Rev. Lett.* **81**, 4863 (1998).
8. Z. Yoshida and S.M. Mahajan, *Phys. Rev. Lett.* **88**, 095001 (2002).
9. Z. Yoshida, S. Ohsaki, A. Ito and S.M. Mahajan, *J. Math. Phys.* **44**, 2168 (2003).
10. Z. Yoshida, S.M. Mahajan and S. Ohsaki, *Phys. Plasmas* **11**, 3660 (2004).
11. Z. Yoshida and Y. Giga, *Math. Z.* **204**, 235 (1990).

SECTION IV

NANOPHYSICS AND APPLICATIONS

Symmetry and novelty in the electronic and geometric structure of nanoalloys: the case of $Ag_{27}Cu_7$

M. Alcántara Ortigoza and T. S. Rahman*

Department of Physics, University of Central Florida
Orlando, Florida 32816/2385, USA
E-mail: alcantar@physics.ucf.edu
** E-mail: talat@physics.ucf.edu*

Nanoparticles of bimetallic alloys have been shown to possess composition dependent characteristics which distinguish themselves from the corresponding bulk alloys. Taking the 34-atom nanoalloy of Ag and Cu ($Ag_{27}Cu_7$), we show using first principles electronic structure calculations that this core-shell alloy indeed has perfect D_{5h} symmetry and consists of only 6 non-equivalent (2 Cu and 4 Ag) atoms. Analysis of the interatomic bond lengths and detailed electronic structure further reveal that the Cu atoms play a major role in controlling the characteristics of the nanoalloy. The higher cohesive energy, together with shorter bond length for Cu, compared to Ag, conspire to produce a hierarchy in the relative strengths of the $Ag - Cu$, $Ag - Ag$, and $Cu - Cu$ bonds and corresponding interatomic bond lengths, point to the uniqueness in the characteristics of this nanoalloy. Charge density plots of $Ag_{27}Cu_7$ provide further insights into the relative strengths of the various interatomic bonds.

Keywords: Bimetallic nanoalloys, Nanoclusters, Alloys lattice dynamics, Electronic structure, charge density

1. Introduction

Metallic nanoalloys, namely multi-component metallic particles less than 10 nm in diameter, have recently received considerable attention because of their unique properties which are of importance not only for fundamental research but also for a range of technological applications. [1] Since these properties are size and composition, hence structure, dependent, such materials provide excellent opportunities for tuning their characteristics by varying these measures. Not surprisingly, nanoalloy properties may be very different from that of the corresponding bulk alloy (there are examples of metals that do not mix in the bulk material but mix at the nanoscale [2]). These properties generally differ also from those of single elements nanoparticles of the same size. [3–5] Of interest are also the attributes of nanocluster-assembled materials whose mechanical properties range from reduced density, to nanoporosity, large surface/volume ratio, and reduced dislocation activity. The magnetic properties range from giant magneto-resistance to superparamagnetic behavior. [6–8] These properties are present already in single-component cluster-assembled materials. The possibility of assembling nanoalloy clusters would open a wealth of new possibilities, since it would make possible to tune not only the mechanical properties (as in the case of pure materials) but also the magnetic and optical ones. This would afford the possibility of exploiting the magnetic anisotropy of some nanoalloy clusters for magnetic storage in cluster arrays. [9,10] This opens up possibilities for a wealth of applications, including quantum dots, quantum computers and devices, data storage devices, nanoscale ferromagnets, optoelectronic devices, chemical sensors, heterogeneous catalysis, etc. [10–14] On the other hand, arrays of core-shell nanoparticles can be used for optical data storage by means of laser-induced interdiffusion within the particles. [12,15,16] Applications in optics include the use of surface plasmons as wave guides on length scales smaller than the optical wavelength. Also, the assembly of multicomponent clusters would open the possibility of searching for new phases and properties at the nanoscale that can be transferred to macroscopic by assembling clusters.

Computational studies are beginning to shed some light on the structure and properties of nanoalloys and their assemblies, [1,3,5,17–23] but a coherent strategy for the computer-aided design of these systems is still lacking. Indeed, the computational study of nanoalloys and their assemblies poses several computational challenges. Concerning isolated nanoalloy clusters in the gas phase, the determination of the most stable structure as a function size and composition is a challenging task. In the cases in which reliable atom-atom interatomic potentials are available or can be developed, one can employ global optimization techniques. [24–26] For binary systems, due to the presence of permutational isomers (homotops) as well as geometrical isomers, the number of possible isomers is much larger that in the case of single-component systems. [1] This limits presently the use of unbiased global optimization techniques to small sizes $N < 100$ atoms. [3,26]

To summarize, until now there has been no unified or systematic attempt to tackle the problem of the computer-aided design of nanoalloys and their assemblies. Such an approach necessitates the integration of several of the most advanced theoretical and computational techniques in materials modeling. The development of this field is thus presently limited by the lack of structural control in the fabrication of nanosystems, i.e., control over the size, size distribution, shape, structure and composition of nanoalloys and of their supra-assemblies. As a consequence, also structure-property relationships also are only poorly understood. Information on these topics is needed urgently, and such information would represent a real breakthrough in nanoscience and nanotechnology. As a first step towards understanding the characteristic of nanoalloys we turn here to the examination of the geometric and electronic structure of a small bimetallic nanoparticle of Ag and Cu for reasons below.

Darby et al. [3], using many-body Gupta potentials, studied the structure and stability (as reflected by the total energy) of a wide variety of Cu_xAu_y nanoclusters with up to 56 atoms and $x/y = 1, 3$; corresponding to the well-known $x : y$ ratios that result in stable ordered bulk phases at low-temperatures. [27] They found that the geometry of the cluster is influenced by the tendency to maximize the number of $Cu - Au$ and $Au - Au$ bonds Rossi et al. [5], on the other hand, proposed a new family of 34-atom bimetallic alloys using the genetic global optimization technique (GGO). [5] These nanoalloys are characterized by a perfect core-shell structure in which the *smaller* atoms (Cu or Ni) compose the core whereas, the relatively larger, Ag atoms lie on the surface. They find the nanoalloys to be energetically and thermodynamically more stable than pure clusters Ni_{34}, Cu_{34}, and Ag_{34}, [5] and attribute the relative stability of the nanoalloy structures to the supplanting of the inner Ag atoms by *smaller* atoms (Cu or Ni) thereby reducing the internal strain in Ag_{34}, or the replacement of outer Cu atoms by *larger* atoms (Ag) to reduce the external strain in Cu_{34}. As we shall see later, the most stable of the 34-atom $Ag - Cu$ nanoalloy family proposed by Rossi et al, [5] $Ag_{27}Cu_7$, provides a hint that it is not the $x : y$ ratio that guarantees the stability of either bulk or nano-alloys, rather it is the maximization of the number of *optimized* $Cu - Cu$ and $Cu - Ag$ bonds.

Among a set of possible core-shell nanoclusters modelled by many-body interatomic potentials, Rossi et al. [5] chose the compositions corresponding to their most stable structures for some selected sizes and *locally* optimized the structures using density functional theory (DFT) to confirm the trends given by the GGO and to single out the clusters with high electronic stability, whose measure they look to be the width of HOMO-LUMO (Highest Occupied Molecular Orbital - Lowest Unoccupied Molecular Orbital) gap. They also determined the thermodynamic stability of the chosen structures through calculations of the melting temperatures from molecular dynamics simulations and of temperature-dependent probabilities of the global minima by harmonic thermodynamics. [28] Among the 34-atom family of nanoclusters, they found $Ag_{27}Cu_7$ and $Ag_{27}Ni_7$ to have the least excess energy with respect to bulk atoms (the lowest heat of formation), strong electronic stability (large HOMO-LUMO gap), and relatively high melting temperatures. [5] For these particular clusters they have proposed a structure with D_{5h} symmetry in which the 7 core-Cu atoms form a decahedral structure while the 27 shell-Ag atoms are placed in an anti-Mackay overlayer.

The purpose of this paper is to examine the relationship between the geometric and electronic structure of $Ag_{27}Cu_7$ nanoalloy to get insights into some of the various factors that may impact its stability. Of interest here will be the formation energy, the density of states near the Fermi level, the HOMO-LUMO gap, and the charge density distribution, from which we expect to develop criteria for the chemical and electronic stability of $Ag_{27}Cu_7$. The rest of the paper is organized as follows: Section 2 contains the computational details, while Section 3 is a summary of our results and is divided in subsections 3.1 and 3.2. Subsection 3.1 analyzes the geometry and bond coordination of the atoms in the $Ag_{27}Cu_7$ nanoalloy. In Subsection 3.2, we examine our calculated DOS of $Ag_{27}Cu_7$ and its charge density distribution. Finally, in Section 4 we summarize our conclusions, and discuss how the relation between structure and stability in $Ag_{27}Cu_7$ nanoalloy can be understood in terms of a specific hierarchy in bond strength and the capability of the structure to provide the bond lengths for which that hierarchy is satisfied.

2. Computational Details

Periodic super-cell calculations are performed in the framework of density functional theory. [29] Our calculations are based on the pseudopotential approach and the plane wave method (Quantum ESPRESSO: opEn-Source Package for Research in Electronic Structure, Simulation, and Optimization). [30] Ultra-soft pseudo-potentials [31] used here are generated consistently with GGA schemes. For the GGA functional the expression introduced by Perdew, Burke, and Ernzerhof (PBE) has been introduced. [32] Integrations up to the Fermi surface are performed by using a broadening technique [33] with smearing parameter of 0.2 eV (0.147 Ry). Since in the unrelaxed configuration of $Ag_{27}Cu_7$ the separation between the most distant atoms was about 8.7 Å, we locate the nanoalloy inside a cubic super-cell with side length of 24 Å. In this manner we ensure that as a result of periodic boundary conditions, the atoms at the edges of neighboring clusters are at least 15 Å apart, thereby isolating the clusters from each other. The same cubic box is used to calculate the total energy of isolated Cu and Ag atoms, using a spin-polarized calculation. The Broyden-Fletcher-Goldfarb-Shanno (BFGS) algorithm [34] is used to minimize the nanoalloy total energy as a function of atomic positions. [30] At equilibrium, forces on the nanocluster atoms are required to be below 6.5×10^{-4} eV/A (2.6×10^{-5} Ry/au). Given the large dimensions of the super-cell used for the nanoalloy and the free atom calculations, integrations over the BZ can be accurately performed using only 1 k-point. To confirm this, we have tested the total energy convergence using 24 special k-points to find that the total energy changes by only 4×10^{-5} eV while forces on each atom remain below 8×10^{-3} eV/A. For the nanoalloy, E_{cut} and E_ρ are 680 eV and 8160 eV, respectively.

3. Results and Discussion

Since the nanoalloy presents several opportunities for comparison of its properties and development of criteria for its stability, we find it beneficial to divide this section into two subsections, each of which consists of several parts. In subsection 3.1, we concentrate on issues related to the geometry, the distribution of bond lengths, the atomic coordination, and the formation energy of the nanoalloy. We first introduce a notation in subsection 3.1.1 that classifies the atoms in $Ag_{27}Cu_7$ according to their location within the nanoalloy. In 3.1.2, we inspect how the local coordination of the atoms in the nanoalloy relates to the bond lengths. The average formation energy of $Ag_{27}Cu_7$ nanoalloy and stability considerations are analyzed in subsection 3.1.3. In subsection 3.2 we focus on the electronic DOS and the local charge density. Subsection 3.2.1 contains the electronic DOS of the $Ag_{27}Cu_7$ nanoalloy and discusses how it compares with those of bulk alloys, Ag_3Cu and Cu_3Ag. [35] The local charge density distribution in the nanoalloy is summarized in subsection 3.2.2.

3.1. *Geometry, bond coordination, and stability considerations*

3.1.1. *Geometric Structure*

The initial configuration we adopted for $Ag_{27}Cu_7$ [36] nanoalloy nicely relaxes towards the D_{5h} symmetry after energy minimization, as seen from the plots in Fig. 1. Accordingly, there are only 6 types of non-equivalent atoms: 4 types of Ag and 2 types of Cu atoms. This leads to a natural and useful classification of the atoms that refers to their distance from the mirror plane: Cu layer 0 ($Cu0$), Ag layer 0 ($Ag0$), Cu layer ±1 ($Cu1$), Ag layer

±2 ($Ag2$), Ag layer ±3 ($Ag3$), Ag layer ±4 ($Ag4$), as shown in Fig. 1. Layer 0, which lies on the mirror plane, consists of two pentagonal structures; the smaller one is made of copper atoms ($Cu0$) and fits in the larger one that is composed of silver atoms ($Ag0$) (Fig. 1(a) and (b)). The *single-atom layers*, layers ±1 and ±4, sit on the 5-fold rotation axis (see Fig. 1(c) and (f)). The other two pentagonal layers, ±2 and ±3, are centered at the 5-fold symmetry axis (Fig. 1(d) and (e)). The radii of $Ag2$ and $Cu0$ pentagons are parallel to each other, but rotated 36° with respect to the $Ag0$ and $Ag3$ pentagons, as shown in Fig. 2. Since the layers are symmetric with respect to a mirror plane, the cluster can be characterized by only five of them, say, the central layer (layer 0) and those above this (layers 1 to 4). Ultimately, the symmetry of $Ag_{27}Cu_7$ allows us to fully describe its geometric structure by 8 parameters: the interlayer distances (d_{01} = 1.341, d_{12} = 0.052, d_{23} = 0.992, d_{34} = 1.512 Å) and the pentagons. side length (a_{Cu0} = 2.584, a_{Ag0} = 4.897, a_{Ag2} = 5.115, a_{Ag3}= 2.948 Å, as shown in Fig. 2).

3.1.2. *Nearest Neighbor Bond Lengths*

Despite the perfect D_{5h} symmetry of the cluster, we find an intricate hierarchy of nearest neighbors (NN) in the optimized structure. For example, for $Ag3$ the 1st NN atoms ($Cu1$) are at 2.726 Å while their 6th nearest neighbors are at 2.948 Å, all within a separation of 3.0 Å. Figs. 3 (a) and (b) show a comparison between the bond coordination of atoms in the $Ag_{27}Cu_7$ nanoalloy and that of atoms in bulk Ag and bulk Cu. The local coordination of Cu atoms in $Ag_{27}Cu_7$ seems, in fact, not dramatically different from that in bulk Cu regarding the number of first and second NN. The $Cu0$ atoms, for example, have 12 neighbors within a distance of 2.548 and 2.786 Å, and another twelve between 4.181 and 4.482 Å. Silver atoms, on the other hand, find themselves in unusual atomic environments: at distances 3 Å (bulk Ag nearest neighbors distance) $Ag2$ and $Ag4$ atoms have barely acquired 6 neighbors, while $Ag3$ and $Ag0$ atoms get 9 and 8 NN, respectively. Between 4.2-4.5 Å, at which bulk Ag atoms already have 18 NN, $Ag0$, $Ag3$ and $Ag4$ atoms have acquired only 11 neighbors, while $Ag2$ atoms have only 7 NN, finding themselves as the most under-coordinated atoms of the cluster. Most importantly, we notice that the NN bond lengths between Cu atoms contract by as much as 2% with respect to the value in bulk. [35] Such contraction may be expected for Cu atoms in low coordination environments, such as on the surface layer of $Cu(100)$ (inward contraction of around 3 % [37]), and not in $Ag_{27}Cu_7$ because of their high coordination (12) as in bulk Cu.

Interestingly, for the shell Ag atoms, which have much lower local coordination than the core Cu atoms, the $Ag-Ag$ NN bond lengths are only 2.6 % shorter than those in bulk Ag.[35] If coordination alone were a measure, one might have expected the most under-coordinated Ag atoms, $Ag2$ and $Ag4$, to undergo a larger contraction, as found on Ag surfaces.[37] Instead, in $Ag_{27}Cu_7$, $Ag2$ and $Ag4$ form short bonds of about 2.59 Å with their neighboring Cu atoms. These bond lengths are in fact very close to the smallest $Cu-Cu$ bond lengths (2.55 Å) and considerably smaller than the smallest $Ag-Ag$ bond lengths (2.87 Å). Indeed, we notice that in $Ag_{27}Cu_7$ the first NN of every Ag atom is a Cu atom, pointing to the reality that finite sized structures of these elements may not follow straightforwardly the behavior of infinite and/or semi-infinite systems and the relationship between bond coordination number and bond stiffening might be subtler in nanoalloys. Two conclusions may, nevertheless, be drawn from the above: 1) the fact that $Cu-Cu$ bond lengths in $Ag_{27}Cu_7$ contract almost as much as $Ag-Ag$ bond lengths suggests that Cu atoms are more sensitive than Ag atoms to either low coordination, or local geometry, or chemical environment; 2) the low coordination of the Ag atoms appears to be significantly compensated by the formation of short bond lengths with Cu atoms. To discriminate between the above mentioned effects of local coordination, geometry, and environment, the conclusions about bond lengths in the $Ag_{27}Cu_7$ nanoalloy need to be put on firmer grounds through examinations of details of the electronic structure and the charge density distribution, and their implications for nanoalloy stability. We will turn to this in Sections 3.1.3 and 3.2.

3.1.3. *Formation energy*

In Ref. 5, the thermodynamic stability of a given nanoalloy is evaluated via considerations of its melting temperature and its relative energetic stability is established through comparison of the heat of formation. Remarkably, although clusters with increasing binding energies per atom do not necessarily have higher melting temperatures,[38] $Ag_{27}Cu_7$ came out with both the highest melting point and the least heat of formation. In terms of stability and minimum-energy structures, however, it is important to know also the output given by the average formation energy, which measures the dissociation or *cohesive* energy of the nanoalloy, and to analyze the meaning, implications, and scope of these two energetic considerations. Thus, to estimate the average strength of the bonds, we calculate the so-called average *formation energy per atom*, E_{form}, which is defined as,

$$E_{form}(Ag_{N_1}Cu_{N_2}) = \frac{E(Ag_{N_1}Cu_{N_2}) - N_1E(Ag_{free}) - N_2E(Cu_{free})}{N}, \tag{1}$$

where $N = N_1 + N_2$ and $E(Cu_{free})$ and $E(Ag_{free})$ are the energies of isolated Cu and Ag atoms, respectively. We find $E_{form}(Ag_{27}Cu_7)$ to be 2.17 eV. Since no experimental data on this particular binary nanocluster exists, we turn to the formation energy of related systems for comparison. For example, calculated cohesive energy of bulk Ag, Ag_3Cu, Cu_3Ag, and Cu are 2.51, 2.66, 3.06, and 3.34 eV, [35] respectively, implying that $E_{form}(Ag_{27}Cu_7)$ is smaller than all these bulk values. Note that the higher cohesive energy of bulk Ag_3Cu as compared to that of bulk Ag signals a stronger $Ag - Cu$ bond than the usual $Ag - Ag$ one. In fact, from the results presented in Ref. 5, one finds that the average formation energy per atom of the 34-atom family decreases monotonically from ~2.6 to ~2.0 eV as the Ag content increases from 0 to 34. Considerations of formation energy alone would thus imply that in this family of nanoalloys the $Cu - Cu$ and $Cu - Ag$ bonds are stronger than the $Ag - Ag$ bonds and that $Ag_{27}Cu_7$ is not the most stable structure. It is thus surprising that a related quantity, i.e. the heat of formation (defined as in Eq. 1, but substituting $E(Cu_{free})$ and $E(Ag_{free})$ by the cohesive energy of Cu and Ag, respectively), plotted as function of the Ag/Cu ratio in Ref. 5, shows a minimum at this intermediate composition ($Ag_{27}Cu_7$) - a result that stands in contrast to that found in $Ag - Cu$ bulk alloys. [39–41] Nevertheless, we find that the heat of formation of the $Ag_{27}Cu_7$ is 9 times larger than that of the hypothetical Ag_3Cu, [35] but this suggestion of instability might be misleading since the heat of formation not only measures the strength of the bonds but also weighs the energetic cost(gain) of breaking(forming) single element bulk bonds to form(from breaking) binary bonds. In reality, the formation of intricately tailored structures as $Ag_{27}Cu_7$ is not expected to occur simply by melting the parent compounds. Thus measures like heat of formation have to be supplemented by others such as the dynamical stability of the alloy as displayed by its vibrational modes. The heat of formation is perhaps more of an indicator of the life-time of the nanoalloy, say, against clustering and the eventual formation of segregated metallic bulk Cu and Ag, if one ignores the energy barriers needed to actually break all bonds in the nanoalloy.

To estimate the strength of the bonds in the 34-atom nanoalloys and understand what distinguishes $Ag_{27}Cu_7$ in its family of nanoalloys, we turn to Fig. 2 of Ref. 5. Rossi et al. show that as the amount of Cu increases up to ~20 % (starting from Ag_34), the heat of formation is reduced or kept constant, implying that small amounts of Cu atoms immersed among Ag atoms (in the nanoalloys $Ag_{34-n}Cu_n$ as n decreases from 7 to 1) create $Cu-Ag$ and $Cu-Cu$ bonds that are stronger than those in bulk Cu and are able not only to counterbalance the cost of the cohesive energy of the newly added Cu atom, but also to increasingly stabilize the nanoalloy. Fig.2 of Ref. 5 shows also that if the content of Cu increases beyond 7 atoms, the heat of formation increases again, with the implication that the strength of the bonds is not able to compensate the bulk Cu cohesive energy for an additional atom. In conclusion, $Ag_{27}Cu_7$ possesses the composition and geometry that maximizes the number of $Cu-Cu$ and $Cu-Ag$ bonds using the minimum number of Cu atoms.

3.2. *Electronic structure and charge density distribution*

3.2.1. *Electronic Density of States*

To obtain the electronic density of states (DOS) of $Ag_{27}Cu_7$ nanoalloy from *ab initio* electronic structure calculations, the discrete states are broadened using Gaussian functions of width 0.14 eV. The resulting DOS of $Ag_{27}Cu_7$ is shown in Figs. 4(a)-(c). The DOS of hypothetical $L1_2$ $Ag-Cu$ bulk alloys of Ref. 35 is shown in Figs. 4(d) and (e) for comparison. First of all, the HOMO-LUMO gap (Δ in Fig. 4(a)) of $Ag_{27}Cu_7$ is found to be 0.77 eV in the ground state, which is only slightly smaller than that reported by Rossi et al, [5] 0.82 eV. As expected, the s-states have negligible contributions and the displayed structures have mostly d-character between -5.5 and -1.7 eV. Fig. 4(b) shows that the centroid of the nanoalloy valence band red-shifts ~ 1 eV as compared to bulk Ag, and blue-shifts ~ 0.5 eV with respect to bulk Cu. Fig. 4(c) shows that even though the amount of Cu in the $Ag_{27}Cu_7$ nanoalloy is ~ 4 times less than that of Ag, it contributes to shift the centroid of the valence band to lower binding energies. As shown in Figs. 4(c)-(e) and in Refs. 27 and 42, the role of Cu is, in general, to enhance the DOS at the top of the valence band and to shift the centroid to lower binding energies, while the effect of Ag is the opposite. Similar results have been reported for $Au-Pd$ nanoclusters, [43] in which the increasing content of Au on Pd clusters reduces the density of states at the Fermi level. Figs. 4(b) and (c) show in addition that the valence band of the nanoalloy is as broad as that of either pure bulk constituents - a point worth of noticing since atoms in a low coordinated environment generally exhibit a valence band narrowing. [43] Pure Ag clusters and Cu_3Au surfaces, [44,45] for example, have shown this effect. The hybridization of Ag and Cu states in $Ag_{27}Cu_7$ thus compensates the d-band narrowing that each atom undergoes.

In general, the features in the electronic DOS that discriminate stable alloy phases are not yet fully understood. Although it is well accepted that the dip in the DOS at the Fermi level is related to the stability of a particular alloy phase, [39] there is no obvious correlation between the two since stable and ordered $Cu - Au$ alloys present a dip in the DOS while $Au - Ag$ alloys do not. [39] In the former the dip in the DOS changes position, width and depth with composition, and structure, and has been found to be related to the electronic specific heat. [27] Kokko et al. [27] noticed also that the dip is considerably lessened in the layered $CuAuI$ phase (which reduces the $Au - Cu$ bond) with respect to Cu_3Au and Au_3Cu. Also, based on their electronic specific heat calculations, they infer that the dip is even smoother in the disordered phases. The nanoalloy $Ag_{27}Cu_7$ displays hardly any dip while the Ll_2 $Ag - Cu$ alloys [35] display one (compare Figs. 4(a), (d) and (e)) similar to, but broader than, that found in stable $Cu - Au$ alloys. [27] Yet, from considerations of the heat of formation, the $Ag - Cu$ alloys are marked as being immiscible. Furthermore, we find that the DOS of the Ag atoms (Figs. 4(b) and (c)) in $Ag_{27}Cu_7$ resembles to some extent that of bulk Ag, despite being highly under coordinated, while that of the fully coordinated Cu atoms is strikingly different from bulk Cu presumably, since half of its neighbors are Ag atoms.

In Ref. 35, it is found that as the content of Ag and consequently the lattice parameter in $L1_2$ $Ag - Cu$ bulk alloys decrease, an increasingly sharper peak is generated at the center of the d-band of Ag atoms since the bottom of the d-band remains almost unchanged and most states from the top of the bulk Ag band retract to lower energies. Despite the later feature, a few states appear above the range of the DOS of bulk Ag, and hybridize with Cu states. The appearance of the higher energy states was related purely to the presence of Cu, whereas the depletion of the top of the Ag d-band is not only due to the presence of Cu but also to the shortening of $Ag - Ag$ bonds at distances smaller than the bond-length of bulk Ag or of $Ag - Cu$ bonds. Low coordination effects, such as depletion the bottom of the DOS and enhancement of the highest peaks of the DOS, were also reported in Ref. 35 for a free standing (111) monolayer. From the above, we conclude that the low coordination of Ag atoms in our $Ag_{27}Cu_7$ nanoalloy can account for the depletion of the bottom of the d-band. On the other hand, noting that the shortest $Ag - Cu$ bond lengths in $Ag_{27}Cu_7$ (around 2.6 Å) are shorter than the bond-lengths in bulk alloys Ag_3Cu

and Cu_3Ag, [35] suggests that the reduction of the DOS at the top of the d-band of $Ag0$ (Fig. 5(a)), $Ag3$ (Fig. 5(c)), and $Ag4$ (Fig. 5(d)) atoms occurs because the presence of Cu at such short distances overpowers their low coordination and no enhancement of the DOS at the top of the Ag d-band occurs, nevertheless, as seen in bulk alloys, the hybridization of Cu and Ag states is at the same time improved by Cu by setting off the occupation of states above the top of the d-band of bulk Ag (see Figs. 5(a)-(d)).

We now turn to the issue of the DOS of Cu atoms in the $Ag_{27}Cu_7$ nanoalloy which differ prominently from that of bulk Cu. We consider three different aspects that may influence their electronic structure: the conspicuously disparate overall geometry, the presence of Ag, and the existence of bond lengths longer than that of bulk Cu. It has been noted [35] that if $Cu - Ag$ bonds are longer than that of bulk Cu, then the bottom of the d-band of Cu is strongly depleted and an increasingly sharper peak at the top is created, which slightly shifts towards higher energies. Similar to the case of Ag, Cu atoms in a low coordinated environment have shown [35] that the bottom of the Cu d-band (states below ~3.5 and ~ 4.0 eV) is entirely extinguished. Interestingly, the d-band of the free monolayer is sharply localized at the top edge, [35] resembling that of $Cu0$ (Fig. 5(e)) and $Cu1$ (Fig. 5(f)) atoms notwithstanding their full coordination in $Ag_{27}Cu_7$. Contracting $Ag - Cu$ or $Cu - Cu$ bonds in bulk alloys also reduces the DOS at the top and enhances the bottom of the d-band, thus improving the $Ag - Cu$ hybridization. [35] Moreover, the presence of Ag was also found to strongly deplete most of the bottom of the d-band of Cu atoms in hypothetical $Ag - Cu$ alloys [35] but, at the same time, to introduce states below the bottom of the DOS of bulk Cu, which hybridize with Ag. [35] From the above, we conclude first that the sharp peak at the top of the DOS of Cu atoms in $Ag_{27}Cu_7$ nanoalloy can only be accounted for by the relatively long distances (~ 2.7-2.8 Å) between Cu atoms and half of their *nearest neighbors* - all Ag atoms - (see Figs. 3(a)), seemingly leading to a weak interaction of the Cu atoms with those *far-lying* Ag neighbors, as occurs for Cu atoms in bulk Ag_3Cu and Cu_3Ag alloys. [35] Second, the effect of the Ag environment on the DOS of Cu atoms in $Ag_{27}Cu_7$ is connected with the quenching of the bottom of the d-band of Cu atoms. We note in addition that the DOS of $Ag_{27}Cu_7$ below ~ 3.5 eV) is remarkably high, as compared to that of bulk alloys in Ref. 35 - despite the low Cu content - and generates a much stronger hybridization between Cu and Ag states, contrasting that observed in bulk alloys. [35] We thus conclude that the *optimum* $Cu - Ag$ bonding in $Ag_{27}Cu_7$ comes about in terms of the electronic DOS through the shortening of $Ag - Cu$ bonds that allows the hybridization of the d-states of Cu and Ag atoms, suppressing by this means the dip featuring in the DOS of Cu_3Ag and Ag_3Cu alloys.

We remark that although the DOS of Ag atoms is not changed as drastically as that of Cu atoms in bulk $Ag - Cu$ alloys - suggesting that Cu is more sensitive than Ag to the chemical environment -, the vulnerability of Cu to the presence of Ag is intermixed with long bond-length effects. That is, the presence of Cu depletes the top of the Ag d-band as much as Ag depletes the bottom of Cu d-band. However, the short $Ag - Ag$ and $Ag - Cu$ bonds (with respect to bulk Ag) as present in both bulk alloys broaden the d-band of Ag atoms, compensating in this manner the effect of the Cu atoms. On the other hand, the $Cu - Cu$ and $Cu - Ag$ bond lengths induced in both bulk alloys never become shorter than that of bulk Cu, rather it is quite the opposite. As a result, narrowing of the d-band of Cu atoms is triggered, aggravating the depletion caused by the presence of Ag atoms and, hence, exaggerating the actual chemical susceptibility of Cu to the Ag environment. In summary, the electronic DOS is found unambiguously related to the bond lengths held in a particular geometry.

3.2.2. *Charge density distribution*

The first aspect that comes to mind in Figs. 6 and 7 of the plotted charge densities of the nanoalloy $Ag_{27}Cu_7$ is that Ag atoms barely supply charge to the surface of the nanoalloy. The question is whether the surface charge depletion coincides with charge redistribution from Ag atoms to Cu atoms since two neighboring metals with significantly different work function can give rise to electron transfer from one metal to the other, as reported in calculations of Pd clusters on $Au(111)$ by Sánchez et al, [45] in correspondence with a work function difference of ~ 0.3 eV. Here, the work function of Cu is larger than that of Ag by ~ 0.2 eV. Indeed, Figs. 6 and 7 display higher charge density around Cu atoms than around Ag atoms.

The plot of the charge density in a plane passing through 10 Ag and Cu atoms in Fig. 6 illustrates that the $Cu0$-$Ag2$ and $Cu1$-$Ag4$ bonds are linked by the highest bonding charge density, corresponding indeed to the shortest $Ag - Cu$ bond lengths (~ 2.58 Å), and implying that they are stronger than the $Cu1$-$Cu1$ (2.58 Å), $Cu0$-$Cu0$ (2.58 Å) bonds and even the $Cu0$-$Cu1$ bonds (2.55 Å), in that order (see Figs. 6 and 7). The next strength of bonding charge density occurs for $Cu0$-$Ag0$ bonds, followed by $Cu1$-$Ag3$ bonds, as shown in Figs. 6 and 7. In this case, the bond lengths are 2.72 and 2.73 Å, respectively. Notice that, in the second set of bonds, the charge density is considerably lower than in $Cu0$-$Ag2$ and $Cu1$-$Ag4$ bonds, which supports the assumption that $Ag - Cu$ interactions die out very rapidly (see Section 3.2.1). The third place in bonding charge density corresponds to $Cu0$-$Ag3$ bonds (not shown) and the shortest $Ag - Ag$ bonds: $Ag0$-$Ag3$ and $Ag2$-$Ag2$ bonds (Fig. 6) whose bond lengths are 2.79, 2.87 and 2.89 Å, respectively. The first might influence very little $Cu0$ atoms since the bond is quite large. However, this bonding charge density appears as large as that of $Ag0$-$Ag3$, which is 0.1 Å further apart. The next in bonding charge density is the $Ag2$-$Ag3$ bond (not shown), whose length is 2.90 Å, followed by the $Ag3$-$Ag4$ bond (Fig. 6), whose length is 2.93 Å. The latter bond length is close to that of the $Ag0$-$Ag3$ bond (and almost identical to the bond length of bulk Ag); however, the charge density bridging these atoms is slightly weaker. We note that $Ag3$-$Ag3$ bonds (not shown) are only slightly longer (2.95 Å) but the bonding charge density around it is less than that around the $Ag3$-$Ag4$ bonds. The charge density around the $Ag0$-$Ag2$ bond (not shown) is substantially lower than that around $Ag3$-$Ag3$, consistent with a larger bond length, 3.0 Å. The next larger bond lengths are more than $\sim$ 4 Å, which are expected to provide much weaker direct interactions that will not be discussed here.

4. Summary

We have presented DFT calculations of the $Ag_{27}Cu_7$ nanoalloy to understand its structure and relative stability via considerations of its energetics, electronic DOS, and charge density distribution. The local coordination of Cu atoms seems not dramatically different from that in bulk Cu regarding the number of first and second NN, whereas Ag atoms find themselves in a very low-coordinated environment but, in return, form $Ag - Cu$ bonds as short as the shortest $Cu - Cu$ bonds. On the other hand, the electronic structure of the Cu atoms in $Ag_{27}Cu_7$ deviates much more from that of atoms in bulk Cu, as compared to the corresponding case of Ag atoms in this nanoalloy.

From our analysis of the geometric and electronic structure, we conclude that the relative stability of $Ag_{27}Cu_7$, among its nanoalloy family, is the result of the maximization of the number of $Cu - Cu$ and $Cu - Ag$ bonds, using the minimum number of Cu atoms. The core-shell $Ag-Cu$ nanoalloys do not behave differently from $Ag-Cu$ bulk alloys regarding the segregation tendency and the migration of Ag to the surface, as pointed earlier, [46] since $Ag_{27}Cu_7$ is segregated by construction. Furthermore, the segregated structure is the attribute that leads to its relative stability, provided that the core-shell structure allows $Cu - Cu$ and $Cu - Ag$ pairs to approach and form bonds nearly as strong as they can be, at almost no expense of either contracting the typical $Ag - Ag$ bond or (most importantly) stretching the typical $Cu - Cu$ bond.

The HOMO-LUMO gap of $Ag_{27}Cu_7$ is found to be 0.77 eV. The DOS of $Ag_{27}Cu_7$ shows similar features as those of $Ag - Cu$ bulk alloys. We find that the novel features of the DOS of full-coordinated Cu atoms in $Ag_{27}Cu_7$ atoms are caused by the relatively long distance separating Cu atoms from half of its first NN. Short $Ag - Cu$ bond-lengths, on the other hand, improve the hybridization of Cu and Ag states; explaining why the hybridization of Cu and Ag states in the $Ag_{27}Cu_7$ nanoalloy is stronger than in $Ag - Cu$ bulk alloys. The observed differences in electronic DOS between $Ag_{27}Cu_7$ and $L1_2$ alloys arise not only because of low-coordination and geometry differences but mainly because the symmetry enforces long $Cu - Cu$ and $Cu - Ag$ bonds, unlike the situation in $Ag_{27}Cu_7$.

In $Ag_{27}Cu_7$, the charge density along $Ag - Cu$ bonds whose length is ~ 2.6 Å is even larger than that around $Cu - Cu$ bonds, and certainly larger than those bridging $Ag - Ag$ atoms. Nevertheless, $Ag - Cu$ bonds whose length is of the order of that in bulk Ag_3Cu, or even Cu_3Ag, [35] are surrounded by an appreciably quenched charge density, explaining why the DOS of Cu atoms show low coordination features. We infer a hierarchy of bond strength, $\{Cu - Cu$, $Cu - Ag$, and $Ag - Ag\}$, correlated to a bond length order, $\{Ag - Ag$, $Cu - Ag$, and $Cu - Cu\}$ so that, the actual strength of the bonds in a particular structure becomes the signature of a relaxed geometry, which may or may not satisfy the above hierarchy, let the strongest bonds to realize, and thus be stable.

Acknowledgments

We are indebted to Prof. Riccardo Ferrando for providing us with the initial configuration of the $Ag_{27}Cu_7$ nanoalloy. This work was supported in part by DOE under grant DE-FG02-03ER46058.

FIG. 1. Three-dimensional picture of the six types of non-equivalent atoms. The red arrow represents the 5-fold rotation axis. (a) Five $Cu0$ atoms sitting on the mirror plane of the cluster, z_0 = 3.88 Å. Note that the z_i-coordinates are given with respect to the reference frame used in the figure and will define later interlayer distances; (b) The five $Ag0$ atoms also sit on the mirror plane, z_0 = 3.88 Å; (c) Two $Cu1$ atoms sitting on the 5-fold rotation axis symmetrically located above and below the mirror plane, z_1= 5.17 Å and z_{-1}=2.59 Å. Notice that $Cu0$-pentagon fits in the $Ag0$-pentagon; (d) Ten $Ag2$ atoms form two pentagons symmetrically located above and below the mirror plane, z_2 = 5.31 Å and z_{-2} = 2.44 Å; (e) Ten $Ag3$ form two pentagons symmetrically located above and below the mirror plane, z_3 = 6.24 Å and z_{-3} = 1.52 Å; (f) The two $Ag4$ atoms sit on the 5-fold rotation axis symmetrically located above and below the mirror plane, z_4 = 7.76 Å and z_{-4} = 0 Å.

FIG. 2. *Top view* of $Ag_{27}Cu_7$ nanoalloy, perpendicular to the mirror plane. The side length of each pentagon as a_{M_n} (see text).

FIG. 3. (a) Bond coordination for Cu atoms of $Ag_{27}Cu_7$ compared with Cu bulk; (b) bond coordination for Ag atoms of $Ag_{27}Cu_7$ compared with Ag bulk.

FIG. 4. (a) Total and projected electronic DOS of $Ag_{27}Cu_7$. The later corresponds to the contribution of s and d atomic states between -7 and 4.2 eV from the Fermi level, E_F, which is set equal to 0. The s-contribution is negligible up E_F. The HOMO-LUMO gap in the ground state, Δ = 0.77 eV, is highlighted in red; (b) comparison between the total DOS of $Ag_{27}Cu_7$ and that of Ag and Cu bulk; (c) contribution from the core (Cu) and shell (Ag) atoms to the projected DOS of $Ag_{27}Cu_7$; (d) DOS of Cu_3Ag showing the d-contribution from each species to the total DOS of the alloy in the $L1_2$ phase; (e) DOS of Ag_3Cu showing the d-contribution from each species the total DOS of the alloy in the $L1_2$ phase.

FIG. 5. The PDOS of the six non-equivalent types of atoms in the $Ag_{27}Cu_7$ nanoalloy is compared with the DOS of pure bulk Cu (black lines) and Ag (blue lines). E_F is shifted to 0 for all of them; (a) $Ag0$; (b) $Ag2$; (c) $Ag3$; (d) $Ag4$ (e) $Cu0$; (f) $Cu1$.

FIG.6. 2D Charge density plot at a plane that contains the 5-fold rotation axis of $Ag_{27}Cu_7$ nanoalloy and is, therefore, perpendicular to its mirror plane. Atoms labelled with red color are precisely centered on that plane.

FIG.7. 2D Charge density plot at the mirror plane of $Ag_{27}Cu_7$. Atoms labelled with red color are precisely centered on that plane.

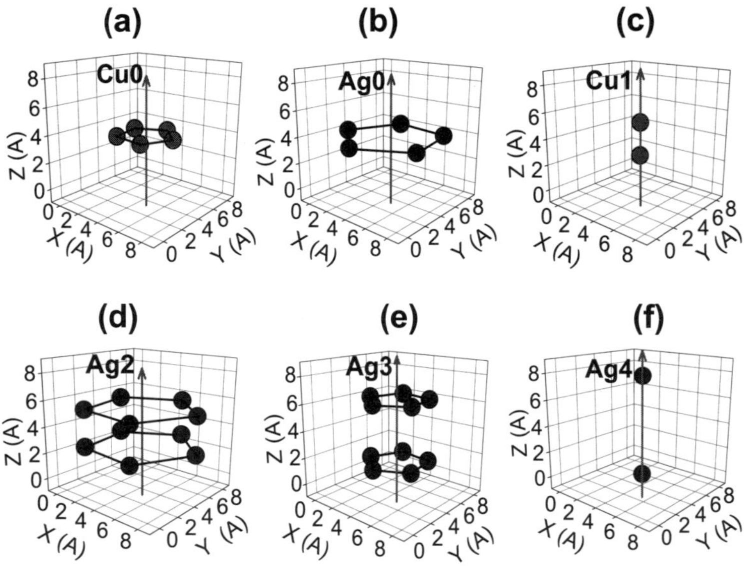
(a)
Cu0
(b)
Ag0
(c)
Cu1
(d)
Ag2
(e)
Ag3
(f)
Ag4
Z (A)
X (A)
Y (A)

Fig. 1.

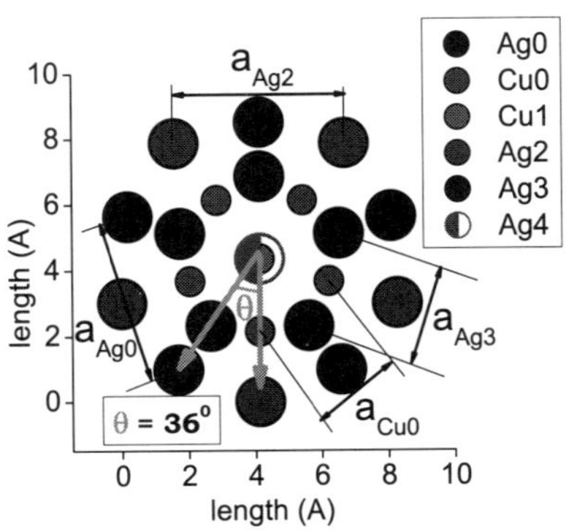
Ag0
Cu0
Cu1
Ag2
Ag3
Ag4
a_{Ag2}
a_{Ag0}
a_{Ag3}
a_{Cu0}
θ
θ = 36°
length (A)
length (A)

Fig. 2.

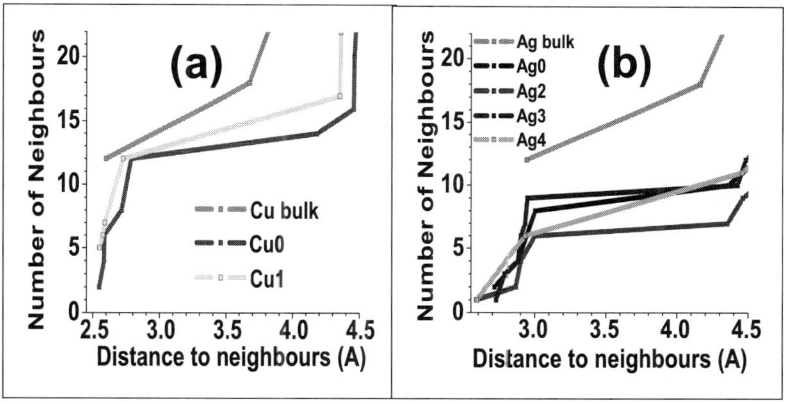
(a)
Number of Neighbours
Distance to neighbours (A)
Cu bulk
Cu0
Cu1
(b)
Ag bulk
Ag0
Ag2
Ag3
Ag4

Fig. 3.

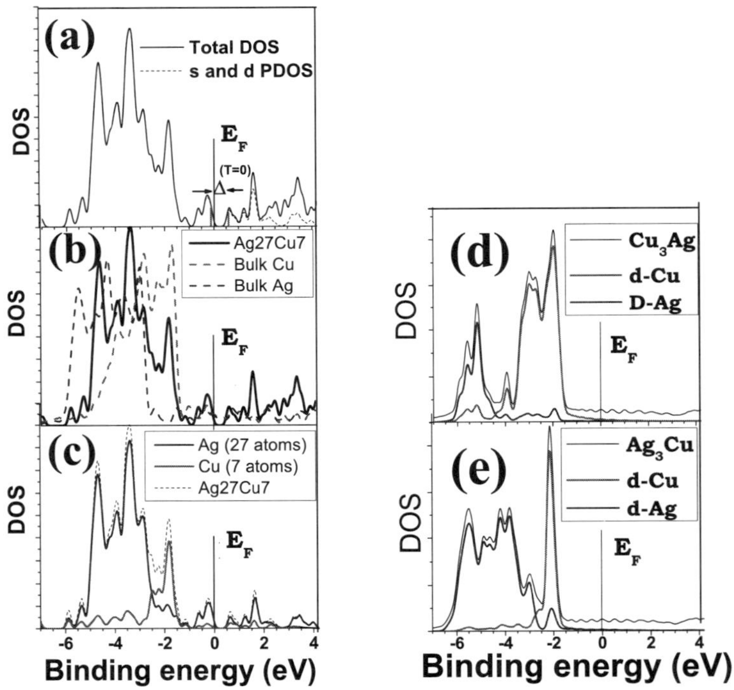
(a)
Total DOS
s and d PDOS
DOS
E_F
(T=0)
Δ
(b)
Ag27Cu7
Bulk Cu
Bulk Ag
(c)
Ag (27 atoms)
Cu (7 atoms)
Ag27Cu7
-6 -4 -2 0 2 4
Binding energy (eV)
(d)
Cu3Ag
d-Cu
D-Ag
(e)
Ag3Cu
d-Cu
d-Ag
Binding energy (eV)

Fig. 4.

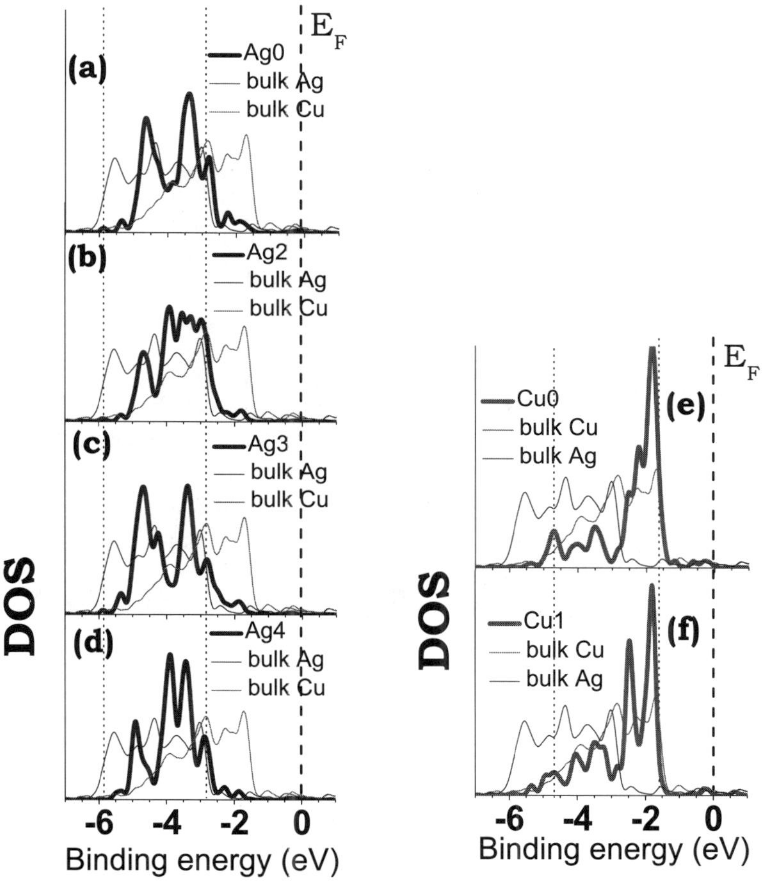

E_F
(a)
Ag0
bulk Ag
bulk Cu
(b)
Ag2
bulk Ag
bulk Cu
(c)
Ag3
bulk Ag
bulk Cu
(d)
Ag4
bulk Ag
bulk Cu
DOS
-6
-4
-2
0
Binding energy (eV)
E_F
Cu0
bulk Cu
bulk Ag
(e)
Cu1
bulk Cu
bulk Ag
(f)
DOS
-6
-4
-2
0
Binding energy (eV)

Fig. 5.

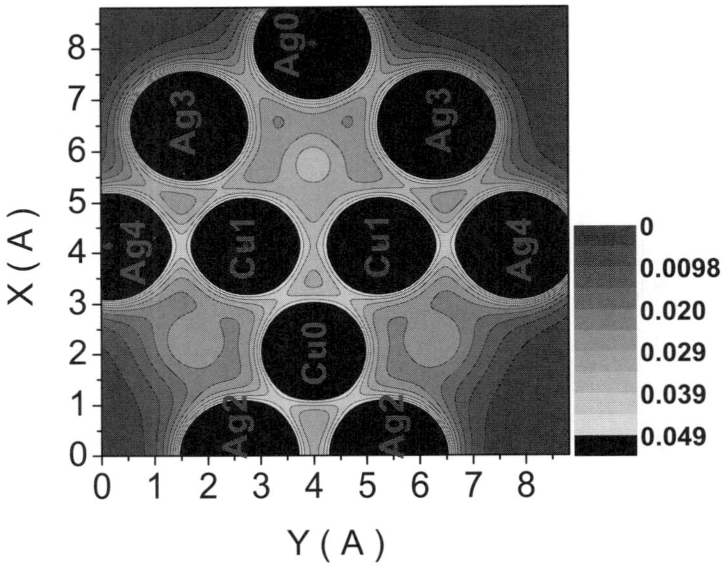
Ag0
Ag3
Ag3
Ag4
Cu1
Cu1
Ag4
Cu0
Ag2
Ag2
X (A)
Y (A)
0
0.0098
0.020
0.029
0.039
0.049

Fig. 6.

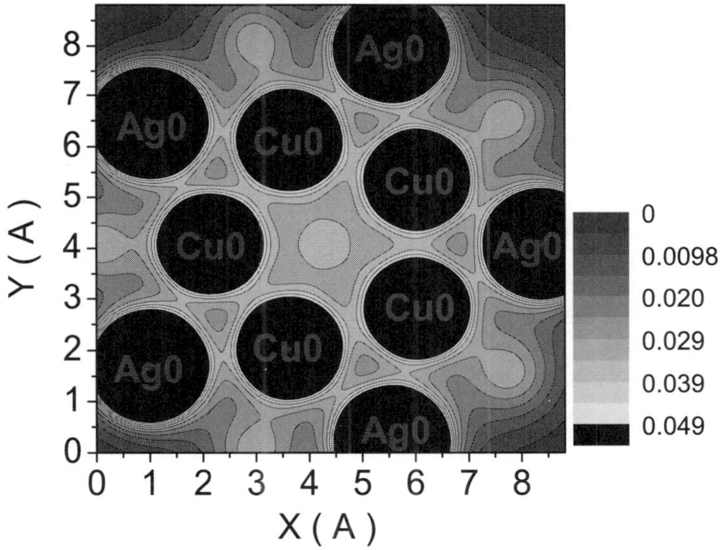
Ag0
Ag0
Cu0
Cu0
Cu0
Ag0
Cu0
Cu0
Ag0
Ag0
0
0.0098
0.020
0.029
0.039
0.049
Y (A)
X (A)

Fig. 7.

References

1. J. Jellinek and E. B. Krissinel, *Theory of Atomic and Molecular Clusters* (J.Jellinek; Springer, Berlin, 1999).
2. M. P. Andrews and S. C. O'Brien, *J. Phys. Chem.* **96**, p. 8233 (1992).
3. S. Darby, T. V. Mortimer-Jones, R. L. Johnston and C. Roberts, *J. Chem. Phys.* **116**, p. 1536 (2002).
4. J. L. Rodríguez-López, J. M. Montejano-Carrizales, U. Pal, J. F. Sánchez-Ramírez, H. E. Trojani, D. García, M. Miki-Yoshida and M. J. José-Yacamán, *Phys. Rev. Lett.* **92**, p. 196102 (2004).
5. G. Rossi, A. Rapallo, C. Mottet, A. Fortunelli, F. Baletto and R. Ferrando, *Phys. Rev. Lett.* **93**, p. 105503 (2004).
6. P. Moriarty, *Rep. Prog. Phys.* **64**, p. 297 (2001).
7. I. Chado, J. P. Bucher, D. Kechrakos and K. N. Trohidou, *J. Phys.: Condens. Matter* **16**, p. S2287 (2004).
8. C. Binns, M. J. Maher, Q. A. Pankhurst, D. Kechrakos and K. N. Trohidou, *Phys. Rev. B* **66**, p. 184413 (2002).
9. C. Petit, S. Rusponi and H. Brune, *J. Appl. Phys.* **95**, p. 4251 (2004).
10. S. Sun, C. B. Murray, D. Weller, L. Folks and A. Moser, *Science* **287**, p. 1989 (2000).
11. J. P. Wilcoxon and P. P. Provencio, *J. Am. Chem. Soc.* **126**, p. 6402 (2004).
12. J. H. Hodak, A. Henglein, M. Giersig and G. V. Hartland, *J. Phys. Chem, B* **104**, p. 11798 (2000).
13. A. M. Molenbroek, J. K. Norskov and B. S. Clausen, *J. Phys. Chem B* **105**, p. 5450 (2001).
14. L. Cao, L. Tong, P. Diao, T. Zhu and Z. Liu, *Chem. Mater.* **16**, p. 6239 (2004).
15. H. Yasuda and H. Mori, *Phys. Rev. Lett.* **67**, p. 3747 (1992).
16. J. I. Park, N. J. Kang, S. M. Lee, S. Kim, S. J. Oh, H. C. Ri and J. Cheon, *Mat. Res. Soc. Symp. Proc.* **721**, p. E5.8 (2002).
17. B. Wang, X. Chen, G. Chen, G. Wang and J. Zhao, *Surf. Rev. Lett.* **11**, p. 15 (2004).
18. T. Van Hoof and M. Hou, *Eur. Phys. J. D* **29**, p. 33 (2004).
19. L. Rubinovich and M. Polak, *Phys. Rev. B* **69**, p. 156405 (2004).
20. C. Mottet, G. Treglia and B. Legrand, *Phys. Rev. B* **66**, p. 045413 (2002).
21. C. Massen, T. V. Mortimer-Jones and R. L. Johnston, *J. Chem. Soc.* **Dalton Trans.**, p. 4375 (2002).
22. F. Baletto, C. Mottet and R. Ferrando, *Phys. Rev. Lett.* **90**, p. 135504 (2003).
23. E. B. Krissinel and J. Jellinek, *Chem. Phys. Lett.* **272**, p. 301 (1997).
24. J. P. K. Doye and D. J. Wales, *Phys. Rev. Lett.* **80**, p. 1357 (1998).
25. J. P. K. Doye and D. J. Wales, *Phys. Rev. Lett.* **86**, p. 5719 (2001).
26. J. Rogan, G. García, C. Loyola, W. Orellana, R. Ramírez and M. Kiwi, *J. Chem. Phys.* **125**, p. 214708 (2006).
27. K. Kokko, E. Ojala and K. Mansikka, *J. Phys. Condens. Matter* **2**, p. 4587 (1990).
28. J. P. K. Doye and F. Calvo, *Phys. Rev. Lett.* **86**, p. 3570 (2001).
29. R. O. Jones and O. Gunnarson, *Rev. Mod. Phys.* **61**, p. 689 (1989).

30. S. Baroni, A. D. Corso, S. de Gironcoli, P. Giannozzi *et al.*, Complete Quantum ESPRESSO distribution version 3.0 (2005), http://www.pwscf.org
31. D. Vanderbilt, *Phys. Rev. B* **41**, p. 7892 (1990).
32. J. P. Perdew, K. Burke and M. Ernzerhof, *Phys. Rev. Lett.* **77**, p. 3865 (1996).
33. M. Methfessel and A. T. Paxton, *Phys. Rev. B* **40**, p. 3616 (1989).
34. K. H. Hellwege and A. M. Hellwege, *Numerical data and functional relationships in science and technology* (K. H. Hellwege and A. M. Hellwege; Landolt-Börnstein-Group III Condensed Matter; New Series; Group III; Springer-Verlag; Berlin, 1969).
35. M. Alcántara Ortigoza and T. S. Rahman, First principles calculations of the electronic and geometric structure of $Ag_{27}Cu7$ nanoalloy, (in progress).
36. R. Ferrando (Private Communication).
37. R. Heid and K. P. Bohnen, *Phys. Rep.* **387**, p. 151 (2003).
38. M. Schmidt, R. Kusche, B. von Issendorff and H. Haberland, *Nature* **393**, p. 238 (1998).
39. K. Terakura, T. Oguchi, T. Morí and K. Watanabe, *Phys. Rev. B* **35**, p. 2169 (1987).
40. J. M. Sánchez, J. P. Stark and V. L. Moruzzi, *Phys. Rev. B* **44**, p. 5411 (1991).
41. A. van-de Walle and G. Ceder, *Rev. Mod. Phys* **74**, p. 11 (2002).
42. K. Kokko, *J. Phys. Condens. Matter* **11**, p. 6685 (1999).
43. G. K. Wertheim, S. B. DiCenzo and D. N. E. Buchanan, *Phys. Rev. B* **33**, p. 5384 (1986).
44. S. B. DiCenzo, P. H. Citrin, E. H. Hartford-Jr and G. K. Wertheim, *Phys. Rev. B* **34**, p. 1343 (1986).
45. C. G. Sánchez, E. P. M. Leiva and W. Schmickler, *Electrochem. Comm.* **5**, p. 584 (2003).
46. F. Baletto, C. Mottet and R. Ferrando, *Eur. Phys. J. D* **24**, p. 233 (2003).

New approaches to photovoltaic and photoelectrochemical energy conversion*

S. Ismat Shah,[1,2] Hong-Ying Lin,[3] Yinghong Miao,[1] Meghan E. Schulz[1†]

1 Department of Materials Science & Engineering,

2 Department of Physics & Astronomy,

3 Department of Civil & Environmental Engineering,

University of Delaware, Newark, Delaware, 19716, USA

In response to skyrocketing fuel costs and evidence of climate change, real technological progress is needed towards harnessing the clean power from the sun to drive human progress. Here we present two approaches to harnessing solar energy currently under investigation in our group. One involves novel photovoltaic cells using different sized quantum dots. Another uses proven photocatalysts to directly electrolyze water, producing hydrogen. The technological background will be discussed, as well as current state of the art and future research direction.

1. Introduction

"Civilized" societies have come to depend on cheap, convenient fossil fuels and feedstocks to power their industrial lifestyle. The specters of pollution and fuel shortages have lurked behind the scenes since the beginning of the industrial revolution itself, spurring fundamental research in alternative and sustainable energy sources. Our work in the field of solar energy focuses on two technologies in particular. First, we present a novel class of solar cells using TiO_2-Ge nanocomposites. Like traditional Si-based photovoltaics (PV), the aim is to produce electric current from solar radiation, which may be used to do work. In the second part, we present recent advances in photoelectrochemical cells (PECs). Instead of electricity, PECs directly convert radiation into chemical energy. This energy is used to electrolyze water, producing the ultimate clean, renewable, and portable fuel: hydrogen gas.[1]

* This work is supported by the National Science Foundation, grant 0441619 and NIRT 0210284

† Work partially supported by the NSF's IGERT program in Solar Hydrogen at the University of Delaware.

2. Advanced Photovoltaics

2.1. *The Shockley-Queisser limit*

In 1961, Shockley and Queisser showed that the maximum theoretical efficiency of a single solar cell was limited to 33%.[2] The Shockley-Queisser limit is due to the thermalization of all photon energy in excess of the band gap. Electrons produced by high energy photons (for traditional Si, anything greater than its 1.1 eV bandgap) dissipate energy to the lattice before being swept across the depletion region. The lattice heats up and the efficiency of a single band gap solar cell decreases, as shown in Figure 1.

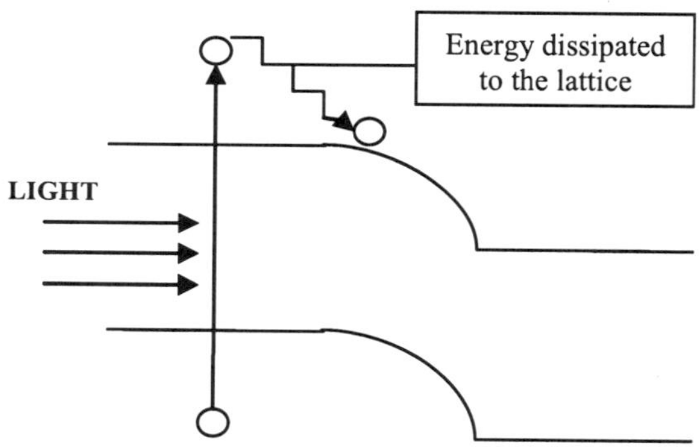

Figure 1. Lattice thermalization process

Two chief approaches to overcoming the Shockley-Queisser limit have been proposed:

1. Increasing the number of bandgaps.
2. Capturing carriers before thermalization.

Several ideas have surfaced with regards to the former technique, increasing the number of bandgaps. Among them are quantum dot sensitized solar cells, multi-junction solar cells (each layer with a different band gap), and quantum wells. We are interested in quantum dot sensitized cells, and the TiO_2-Ge nanocomposite solar cell is used as an example below to explain the working principles and advantages.

2.2. *TiO_2-Ge nanocomposite working principle*

The Bohr radius of germanium is 24.3 nm. The E_g of bulk Ge is 0.66 eV. By tailoring the size of Ge quantum dots, the bandgap of Ge can be tailored from

0.66 up to about 2.2 eV due to the well-known quantum confinement effect. Thus, a wide range of light may potentially be absorbed by Ge quantum dots, from the infrared to the visible.

Our system involves Ge quantum dots embedded in a TiO_2 thin film, as shown in Figure 2; TCO conduction glass and Pt are used as the electrodes. By carefully choosing electrode materials to match the work function, and adding electrolyte, it is possible to collect light-generated carriers.

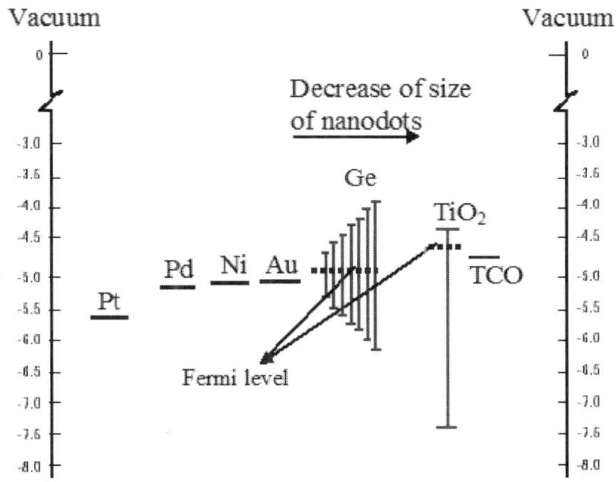

Figure 2. Energy band diagram of TiO_2.

When this system is illuminated, an exciton pair is created in Ge dots. The electron excited to the conduction band of Ge can fall back into the conduction band of TiO_2. The holes are transferred to Pt electrode via an electrolyte added in the solar cell system.

Other kinds of nanocomposite solar cells have been reported in many papers. Examples include InAs sensitized TiO_2 solar cells,[3] and $TiO_2/CuInS_2$ nanocomposite solar cells.[4]

2.3. *Advantages of nanocomposite crystalline solar cells*

Quantum dot based solar cells have the potential to increase the maximum attainable thermodynamic conversion efficiency of solar photon conversion past the Shockley-Quessier limit, up to about 66%, by tailoring the energy gap of quantum dots. Simultaneously, semiconductor nanocrystals have the potential benefit of greatly decreasing cost of cell manufacture.

3. Photocatalytic Water Splitting

3.1. *Mechanism*

The initiation of a photocatalytic reaction requires a minimum photon energy that equals or exceeds the band gap of the material in order to trigger the interband transition of electrons between the lowest unoccupied molecular orbital (LUMO) and the highest occupied molecule orbital (HOMO). That is, the incident wavelength needs to be smaller than or equal to the wavelength of the band gap threshold, λ_{bg}. Thus, it is speculated that reducing the band gap of TiO_2 can enhance its photocatalytic performance through more efficient utilization of lower energy photons.

TiO_2 has three distinct crystalline structures: rutile, anatase, and brookite. Most studies on the photocatalytic reactivity were conducted with either rutile or anatase, which reported that the indirect band gap was 3.0 (or λ_{bg} 413 nm) and 3.2 eV (or λ_{bg} 387 nm), respectively. The band gap of TiO_2 is indirect. In contrast to direct interband transition, indirect transition requires phonons (lattice vibration) to compensate for the change in wave vector in k-space during electron transition.

The principle of photocatalytic reaction in colloidal systems is to oxidize the electron rich solutes (i.e. soluble organic compounds), through either a direct hole oxidation or an indirect oxidation by hydroxyl radical generated at the valence band. In such a reaction scheme, photo-excited electrons are scavenged mainly by dissolved oxygen molecules to complete the electron transfer cycle. Clearly, this leads to a waste of photo-generated electrons.

We would like to use such a photocatalytic system to produce hydrogen from water; hydrogen being a promising sustainable energy resource as mentioned above. For water electrolysis in a semiconductor particulate system, the energy of the conduction band edge must be higher than the upper water stability limit $E(H^+/H_2)$ on the absolute energy scale. This, in fact, is a difficult requirement to meet in a particulate system without appropriate energy alignment between Fermi energy of soluble species and valence/conduction band edges of the semiconductor photocatalyst.

Fujishima and Honda[5] first reported photocurrent generation using TiO_2 semiconductor electrodes in 1972. Their attempt showed the feasibility of conversion of photon energy into chemical energy (H_2) without wasting of photo-induced electron excitation in photocatalytic systems. The photoelectrochemical system (PEC) consists of a photoelectrode and a counter electrode.[1] When electron-hole pairs are generated due to efficient photo excitation, the holes created on the anode will oxidized adsorbed OH^- and discharge the O_2 via an oxidation reaction:

$$2p^+ + 2OH^- \rightarrow \frac{1}{2}O_2 + H_2O \tag{1}$$

Simultaneously, the photo-excited electrons migrate to the cathode through external circuit and generate H_2 via a reduction reaction:

$$2e^- + 2H_2O \rightarrow H_2 + 2OH^- \text{ or} \tag{2}$$

$$2e^- + 2H^+ \rightarrow H_2 \tag{3}$$

The PEC system allows a wide margin of energy alignment within the semiconductor/liquid junction and offers a solution for utilization/storage of these photogenerated electrons depending on the system configuration and setup. The PEC system shares the same principle with Gratzel-type dye-sensitized solar cell (DSSC) except that the dyes in DSSC can be regenerated via charge transfer through I_3^-. As the demand of clean energy sources remains stringent, there is a need for development of economic yet efficient photovoltaic devices. Due to its excellence in electro-optical and thermal dynamic properties, TiO_2 have been recognized as one of the most suitable materials for photon harvesting application.

3.2. *Fundamental limitations of PECs*

The major hindering effects associated with photocatalytic water splitting in PEC system are:

1. Relatively large band gap (e.g. 3.0–3.2 eV for intrinsic TiO_2) limits the quantum conversion efficiency for solar spectrum photons.
2. Materials with optimal band gaps are often unstable when immersed in the aqueous electrolyte of a traditional PEC.

In the next sections, we will give a detailed explanation of how a modified TiO_2 electrode may mitigate these effects.

3.2.1. *Inefficiency*

To improve the quantum conversion efficiency in a TiO_2-based PEC system, the most efficient approach is to reduce the bandgap of the semiconductor. The bandgap of TiO_2 is determined by its electronic structure. Theoretical studies based on first-principles orthogonalized linear-combinations of atomic-orbitals

(OLCAO)[6] and full-potential linearized augmented plane-wave (FLAPW)[7] have revealed that the upper valence band predominately consists of O2*p* states and the minimum of the conduction band is mainly constructed with Ti3*d* states. In principal, if energy states could be introduced within this forbidden energy band, considerable reduction in the bandgap could be obtained. Therefore, quantum efficiency is enhanced by an increase in the visible light photon absorption.

The most promising method to approach this goal is to dope the impurities into TiO_2 lattices.[8-16] The reported substitutional doping of metal[9,11,14,17] and non-metal TiO_2 catalysts[8,10,12-14,18,19] have often been proven to improve the photocatalytic performance of TiO_2. Among all dopants reported, nitrogen doped TiO_2 is consistently recognized as one of the most visible light sensitive photocatalyst, both theoretically and experimentally.[8,13-16,18,20-22] The density of states (DOS) calculation based on spin restricted local density approximation (LDA) from Asahi[8] predicted that substitutionally doped $TiO_{2-x}N_x$ creates localized N2*p* states just above the valance band edge. This results a mild reduction in band gap (~0.3 eV) and red shifts the optical absorption edge around 100 nm beyond the UV light region.

Asahi[8] also showed that the interstitial and mixture of substitutional and interstitial N doped TiO_2 could introduce deep N2*p* states within the forbidden band. Although this significantly reduces the bandgap of TiO_2, it also encourages the rate of recombination and reduces the charge carrier mobility, which could be a hindering effect for photocatalysis. For the soluble organic removal in a colloidal suspension system, the oxidative power not only depends on how low the valence band edge is in absolute energy scale, but it also requires the energy of the conduction band edge to be higher than the bottom of water stability limit (O_2/H_2O) to promote the conduction band electron scavenging and prolonging the electron transfer kinetic. Thus, deep N2*p* localized states created by interstitial doping may give a detrimental effect to photocatalysis.[16]

3.2.2. *Corrosion*

An important criterion for photoanode materials selection is its resistance to corrosion. Many semiconductors with promising electronic properties are unstable in aqueous environments. If the photoanode is itself oxidized by holes at the electrolyte interface, the electrochemical equilibrium is upset.[23] There are two strategies for conquering this obstacle: either select materials resistant to corrosion in water, or to modify the electrolyte itself to prevent corrosion.

We use the former technique: picking or designing corrosion-resistant materials. To understand why some materials react with water, we return to the band structure diagrams.

To optimize the water splitting performance in a PEC system, the semiconductor photoanode is required to have its valance band edge positioned below the lower water stability limit. Several semiconductors that meet this requirement have been reported (Figure 3).[1,24] Among these materials, Fe_2O_3 has the lowest band gap (Eg = 2.3 eV). However, it is unstable in aqueous environments due to corrosion.

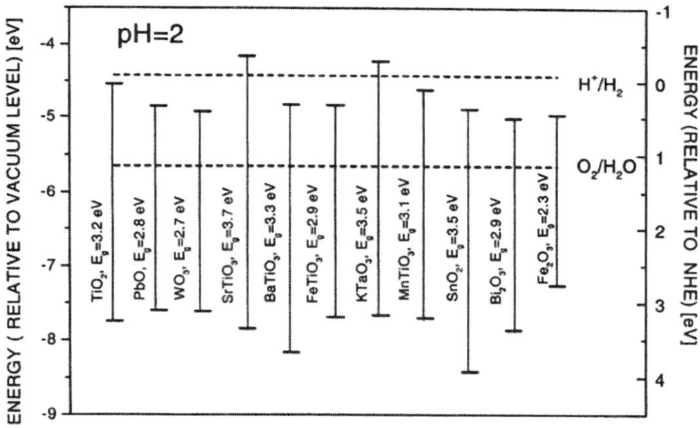

Figure 3. Band gap energy of different semiconductor oxide materials and their band edge positions with respect to the absolute energy scale and normal hydrogen electrode energy in aqueous phase at pH=2.[1,24]

Figure 4 shows in detail the origins of such corrosion problems. According to Gerischer,[25] the rules for electrode stability against corrosion are as follows:

$$E(O_2/H_2O) < E_{p,d} \quad (4)$$

$$E(H^+/H_2) > E_{n,d} \quad (5)$$

Where $E(O_2/H_2O)$ and $E(H^+/H_2)$ are the electrochemical potentials for either half of the electrolysis reaction, $E_{p,d}$ and $E_{n,d}$ are the free energies of oxidation and reduction of the electrode material itself. From Figure 4 we see that GaAs has an advantageous band gap for used solar fraction; however, its decomposition potentials are such that it will easily dissolve in an aqueous electrolyte. On the other hand, SnO_2, WO_3, and TiO_2 are all sufficiently unreactive in water.

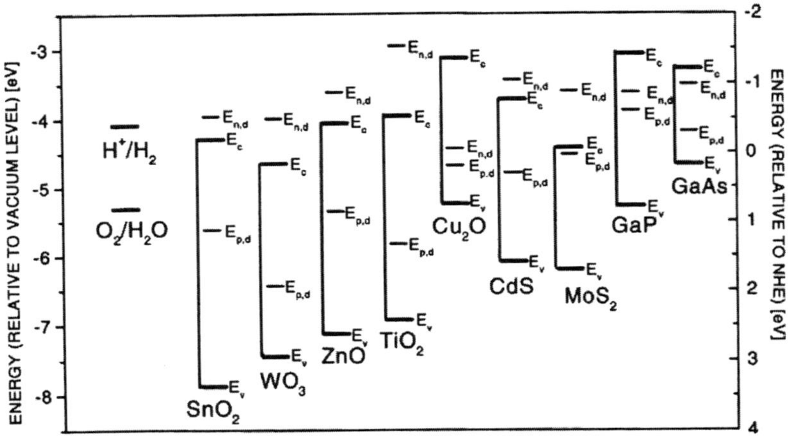

Figure 4. Relative positions of decomposition potentials vs. energy band positions for some semiconducting materials.[1,25]

The most commonly used semiconductor oxides for PEC applications have been reported to be TiO_2 and SrTiO3.1'[5,26] Although TiO_2 has lower band gap energy than that of $SrTiO_3$, it has been reported by Mavroides et. al.[26] that $SrTiO_3$ yields one order higher quantum efficiency in photoelectrolysis comparing to TiO_2 electrode without additional voltage bias condition. They concluded that the difference is due to the lower electron mobility in TiO_2, which is caused by band bending.

There are other possible techniques for protecting the electrode. For example, one may add a thin protective layer for the photoelectrode that would prevent direct contact with the electrolyte, but still allow exchange of light-generated carriers.[24] We could add electronegative species near the interface, to compete with the anode for electrons.[23] Another possibility is a non-aqueous electrolyte, such as an organic solution, or something polymer-based.

3.3. *Strategy for water splitting by PEC systems*

3.3.1. *Energy Alignment*

In solid state physics, the characteristic energy positions [(e.g. Fermi energy E_F, conduction and valance band edges (E_C and E_V)] of a semiconductor material are most commonly expressed on the absolute vacuum energy scale (AVS). However, electrochemistry usually positions the redox couple energy of a specific element with respect to the normal hydrogen electrode (NHE). The absolute vacuum energy scale has been reported to be offset by -4.5 eV from redox energy at normal hydrogen electrode scale at temperature of 25°C.

Figure 5 is a schematic energy diagram of an n-type semiconductor/ electrolyte interface (a good example of the TiO_2/solution junction system). From the figure, we can see both valance and conduction band edges bend upward from its flat band potential due to the presence of depletion region (Schottky barrier) which, in turn, gives a plausible effect if a reduction reaction is desired, but a hindering effect when oxidation reaction is wanted. The direction of electron flow in a semiconductor/liquid junction PEC system is dictated by the relative energy position between the semiconductor band edges (E_C and E_V) and the Fermi energy of the redox couple species ($E_F{}^o{}_{redox}$) at the solid/liquid interface. Hence, the greater difference between the Fermi energy of the redox couple and energy at valance band edge, the higher driving force for promoting oxidation process is created. Using an electron rich electron donor

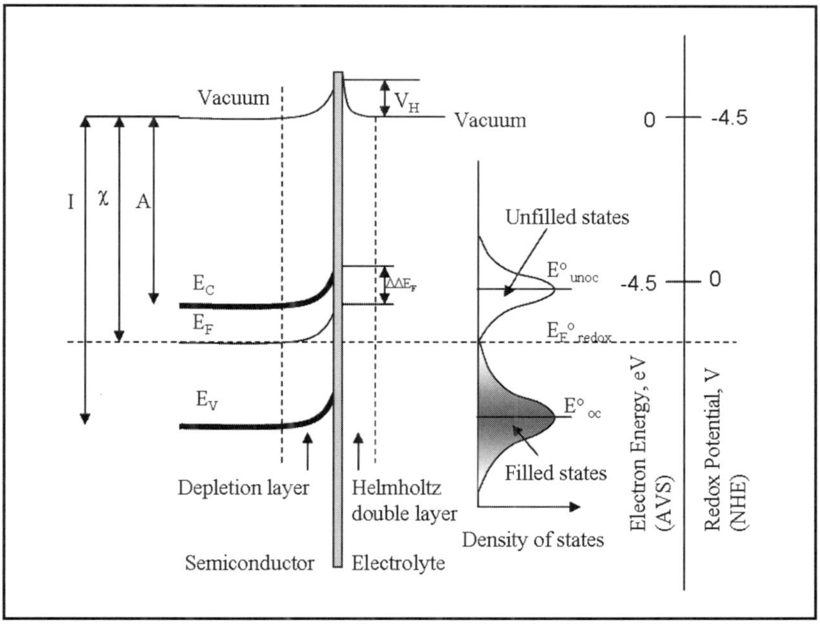

Figure 5. Energy diagram of the interface between an n-type semiconductor surface and electrolyte. The energy scale is presented in both absolute vacuum energy scale (AVS) and its corresponding normal hydrogen electrode scale (NHE). A is the electron affinity, χ is the electron negativity, I is the ionization energy, E_F is the Fermi level energy of the system, E_C and E_V are the energy level at conduction (LUMO) and valance band edge (HOMO), ΔE_F is the increased potential at interface due to band bending, V_H is the potential drop of Helmholtz layer, $E^o{}_{unoc}$ and $E^o{}_{oc}$ are the energy levels of the unoccupied and occupied redox couple. $E_F{}^o{}_{unoc}$ is the Fermi level energy at system equilibrium.

rather than direction oxidation of water molecular is another approach to enhance the performance of a PEC system.

For reduction reaction (hydrogen evolution), a catalyst with a higher redox potential is normally preferred as cathode material. Platinum ($E(Pt^{2+}/Pt) = 1.19$ V vs. NHE) is the most commonly used material in PEC systems. However, the energy difference between redox potential of cathode material and upper water stability limit (H^+/H_2) is fixed at constant pH value. Thus, the only way to enlarge the driving force for the reduction reaction (H_2 generation) is to apply a forward biased potential between the working electrode and counter electrode.

By applying a forward or reversed biased potential, the magnitude of band bending and the direction of electron injection can be maneuvered, yielding a high resolution energy band alignment tenability. The energy alignment between redox species and band edges of semiconductor electrode can be achieved by controlling type and concentration of dopant, material of cathode and photoanode substrate, types of reactant, and direction/magnitude of bias voltage in PEC systems.

3.3.2. *PEC Designs*

As we mentioned above, there are two main factors that affect the quantum conversion efficiency of PEC systems:

1. Band gap of the semiconductor material, which
2. governs the amount of the photon that can be utilized through photoexcitation process. The reduction of semiconductor band gap can be done by doping of impurities that creates impurity electronic states within its intrinsic forbidden band and reduces its effective band gap.
3. Relative energy positions between the redox species and Fermi energy or band edge energy of electrodes. This dictates the direction of electron flow and determines the performance of PEC system.

To maximize the electron driving force between the solid/liquid junction, an appropriate energy alignment is necessary. A schematic design of a PEC cell is shown in Figure 6. This dual chamber designed PEC system uses band gap modified TiO_2 thin film electrode to allow broader photon spectrum harvesting. This design also uses an ion exchange membrane to mediate the charge species which permits only the cations (degraded from large organic molecules) from the anodic chamber (right) to cathodic chamber (left). Filling of the N_2 gas prior to the operation is critical to prevent electron scavenging by oxygen molecular in the cathodic chamber.

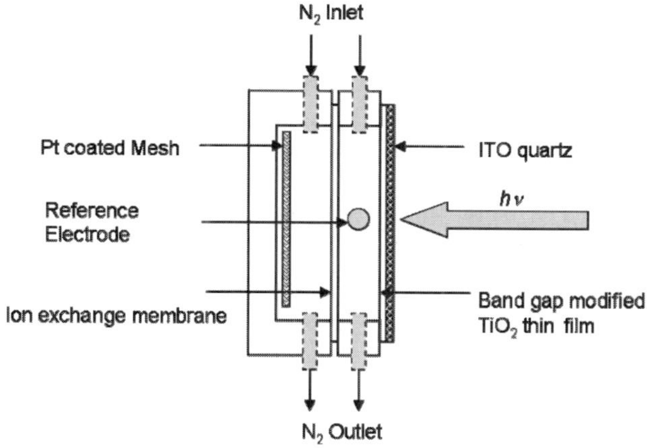

Figure 6. Schematics of a PEC cell

Acknowledgments

The authors wish to acknowledge support from NSF NITR (Grant No. DMR-0210284) and NSF ACT (Grant number DMR-0441619) to carry out this research.

References

1. T. Bak, J. Nowotny, M. Rekas, C. C. Sorrell, *Int. J. Hydrogen Energy* **27,** 991 (2002).
2. W. Shockley and H.J. Queisser, *J. Appl. Phys.* **32,** 510 (1961).
3. P. Yu, K. Zhu, A.G. Norman, S. Ferrere, A.J. Frank, A.J. Nozik, *J. Phys. Chem. B* **110,** 25451 (2006).
4. M. Nanu, J. Schoonman, A. Goossens, *Nano Lett.* **5,** 9 (2005).
5. A. Fujishima and K. Honda, *Nature* **238,** 37 (1972).
6. S.D. Mo and W.Y. Ching, *Phys. Rev. B.* **51,** 13023 (1995).
7. R. Asahi, Y. Taga, W. Mannstadt, A.J. Freeman, *Phys. Rev. B.* **61,** 7459 (2000).
8. R. Asahi, T. Morikawa, T. Ohwaki, K. Aoki, Y. Taga, *Science* **293,** 269 (2001).
9. L. Zang, W. Macyk, C. Lange, W.F. Maier, C. Antonius, D. Meissner, H. Kisch, *Chem. Eur. J.* **6,** 379 (2000).
10. T. Umebayashi, T. Yamaki, H. Itoh, K. Asai, *Appl. Phys. Lett.* **81,** 454 (2002).
11. T. Sumita, T. Yamaki, S. Yamamoto, A. Miyashita, *Thin Solid Films* **416,** 80 (2002).

12. S. Sakthivel and H. Kisch, *Angew. Chem. Int. Ed.* **42,** 4908 (2003).
13. T. Lindgren, J. Lu, A. Hoel, C.G. Granqvist, G.R. Torres, S.E. Lindquist, *Sol. Energy Mater. Sol. Cells* **84,** 145 (2004).
14. S. Z. Chu, S. Inoue, K. Wada, L. Di, J. Suzuki, *Langmuir* **21,** 8035 (2005).
15. C. Di Valentin, G. Pacchioni, A. Selloni, S. Livraghi, E. Giamello, *J. Phys. Chem. B* **109,** 11414 (2005).
16. K.S. Yang, Y. Dai, B.B. Huang, S.H. Han, *J. Phys. Chem. B* **110,** 24011 (2006).
17. T. Umebayashi, T. Yamaki, H. Itoh, K. Asai, *J. Phys. Chem. Solids* **63,** 1909 (2002).
18. R. Nakamura, T. Tanaka, Y. Nakato, *J. Phys. Chem. B* **108,** 10617 (2004).
19. C. Di Valentin, G. Pacchioni, A. Selloni, *Chem. Mater.* **17,** 6656 (2005).
20. G.R. Torres, T. Lindgren, J. Lu, C.G. Granqvist, S.E. Lindquist, *J. Phys. Chem. B* **108,** 5995 (2004).
21. C. Di Valentin, G. Pacchioni, A. Selloni, *Phys. Rev. B: Condens. Matter* **70** (2004).
22. H. Irie, Y. Watanabe, K. Hashimoto, *J. Phys. Chem. B* **107,** 5483 (2003).
23. A.W. Bott, *Current Separations* **17,** 87 (1998).
24. S. Chandra, *Photoelectrochemical Solar Cells*; New York: Gordon and Breach, 1985.
25. J. Gerischer, in: Seraphin, B.O., ed. *Solar Energy Conversion.* Berlin: Springer, 1979
26. J.G. Mavroides, J.A. Kafalas, D.F. Kolesar, *Appl. Phys. Lett.* **28,** 241 (1976).

SECTION V

PARTICLE PHYSICS, GRAVITY AND COSMOLOGY

Theoretical interest in B-meson physics at the B factories, Tevatron and the LHC

Ahmed Ali

Deutsches Elektronen-Synchrotron DESY, Hamburg, Germany
E-mail: ahmed.ali@desy.de

We review the salient features of B-meson physics, with particular emphasis on the measurements carried out at the B-factories and Tevatron, theoretical progress in understanding these measurements in the context of the standard model, and anticipation at the LHC. Topics discussed specifically are the current status of the Cabibbo-Kobayashi-Maskawa matrix, the CP-violating phases, rare radiative and semileptonic decays, and some selected non-leptonic two-body decays of the B mesons.

Keywords: B-Meson Physics; B Factories; Tevatron; LHC

1. Introduction

The two B-meson factories operating at the KEK and SLAC e^+e^- storage rings have outperformed their projected luminosities and have produced a wealth of data on the process $e^+e^- \to \Upsilon(4S) \to B\bar{B}$, with subsequent weak decays of the B and $\bar{B}$ mesons. The exclusivity of the final state, the asymmetric beam energies and the large statistics collected at the $\Upsilon(4S)$ ($O(10^9)B\bar{B}$ events) have led to a number of impressive and quantitative results, which include, among other measurements, precise determination of the weak mixing (Cabibbo-Kobayashi-Maskawa CKM [1,2]) matrix elements $|V_{cb}|$ and $|V_{ub}|$ [3], a large number of CP-violating asymmetries in exclusive decays of the B mesons [4], and rare radiative and semileptonic decays (implying flavor changing neutral current FCNC transitions) involving the B^0 and B^+ mesons and their charge conjugates [5]. Recently, Belle collaboration have published a number of interesting results on the decays of the B_s-meson highlighted by the first measurement of the radiative penguin decay $B_s \to \phi\gamma$ [6]. This is the SU(3)-counterpart to the much celebrated decay $B \to K^*\gamma$, first measured by CLEO and further consolidated by the BABAR and BELLE collaborations [7].

The two B-meson factories are not the only players currently active in the study of B-meson physics. In fact, b-physics is the main beneficiary of the upgrade program at the high energy proton-antiproton collider Tevatron at Fermilab. Since this upgrade, physics news from Fermilab are dominated by the achievements in the b-quark sector. We have seen very impressive and seminal results in this field by the two Fermilab experiments CDF and D0, including the landmark measurement of the B_s^0 - $\overline{B_s^0}$-mixing induced mass difference ΔM_{B_s} [8,9] and the first result on direct CP violation in the B_s-meson sector $\mathcal{A}_{\rm CP}(B_s^0 \to K^+\pi^-)$ [10,11]. These experiments have established (if any proof was needed) that cutting edge flavor physics is done also at hadron machines. In all likelihood, this success story of the hadron machines will be set forth at the Large Hadron Collider (LHC), commissioned to operate in 2008 at CERN. In particular, the experiment LHCb [12], dedicated to precision measurements of the physics of the entire family of B-mesons (B^0, B^+, B_s^0, B_c^+, their charged conjugated states, and excited states) and Λ_b baryons (including the entire baryon spectrum containing at least one b-quark), will greatly broaden our knowledge of both the spectroscopic and dynamical aspects of b-physics. The other two LHC experiments ATLAS and CMS will also contribute to b physics.

In this talk, I will briefly review the highlights of the measurements already accomplished, making contacts with theoretical expectations in the standard model (SM). The topics discussed are: (i) an update of the CKM matrix, with emphasis on the matrix elements in the third row and the third column of $V_{\rm CKM}$ from direct decays (which determine $|V_{cb}|, |V_{ub}|$ and $|V_{tb}|$), and from induced transitions involving the mass differences ΔM_{B_d}, ΔM_{B_s} and the electromagnetic penguin decays $b \to (s,d)\gamma$ (which determine the matrix elements $|V_{td}|$ and $|V_{ts}|$) (ii) measurements of the CP-violating phases α, β and γ from exclusive B-decays, and (iii) radiative, semileptonic and leptonic rare B-decays, a field which has received a lot of theoretical attention both in the context of the SM and in extensions of it. Much more detailed discussions of these topics can be found in the original literature cited below, in the reviews, such as the Review of Particle Properties by the Particle Data Group [13], and in the proceedings of the topical workshops on flavor physics [14].

2. Status of the CKM Matrix

The CKM matrix is written below in the Wolfenstein parameterization [15] in terms of the four parameters A, λ, ρ, η:

$$V_{\rm CKM} \equiv \begin{pmatrix} 1-\frac{1}{2}\lambda^2 & \lambda & A\lambda^3(\rho-i\eta) \\ -\lambda(1+iA^2\lambda^4\eta) & 1-\frac{1}{2}\lambda^2 & A\lambda^2 \\ A\lambda^3(1-\rho-i\eta) & -A\lambda^2\left(1+i\lambda^2\eta\right) & 1 \end{pmatrix} . \qquad (1)$$

Anticipating precision data, a perturbatively improved version [16] of the Wolfenstein parameterization will be used below with $\bar{\rho}=\rho(1-\lambda^2/2)$, $\bar{\eta}=\eta(1-\lambda^2/2)$.

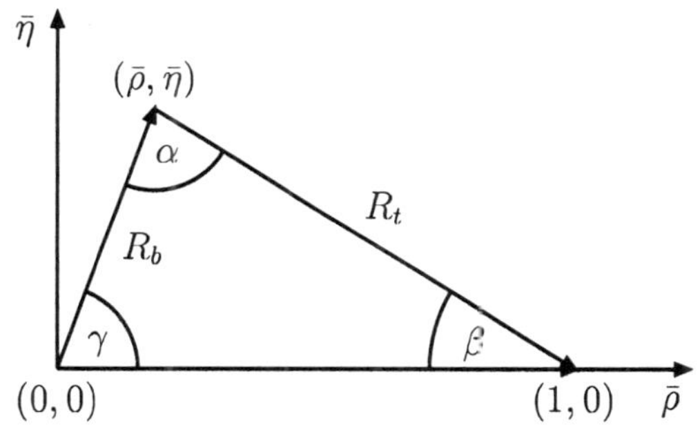

Fig. 1. The unitarity triangle with unit base in the $\bar{\rho}$ - $\bar{\eta}$ plane.

Unitarity of the CKM matrix implies six relations, of which the one resulting from the equation $V_{ud}V_{ub}^* + V_{cd}V_{cb}^* + V_{td}V_{tb}^* = 0$ is the principal focus of the current experiments in B-decays. This is a triangle relation in the complex plane (i.e. $\bar{\rho}$–$\bar{\eta}$ space), and the three angles of this triangle are called α, β and γ, with the BELLE convention being $\phi_1=\beta$, $\phi_2=\alpha$ and $\phi_3=\gamma$ (see Fig.1). The unitarity relation in discussion can also be written as

$$R_b e^{i\gamma} + R_t e^{-i\beta} = 1\,, \qquad (2)$$

where $R_b = \left(1-\frac{\lambda^2}{2}\right)\frac{1}{\lambda}\left|\frac{V_{ub}}{V_{cb}}\right| = \sqrt{\bar{\rho}^2+\bar{\eta}^2}$ and $R_t = \frac{1}{\lambda}\left|\frac{V_{td}}{V_{cb}}\right| = \sqrt{(1-\bar{\rho})^2+\bar{\eta}^2}$. Thus, precise determination of $|V_{cb}|$, $|V_{ub}|$ and $|V_{td}|$ and the three CP-violating phases α, β, γ is crucial in testing the CKM paradigm.

In this section, we mainly review the current status of the CKM matrix elements in the third row and the third column, namely $V_{ub}, V_{cb}, V_{td}, V_{ts}$ and V_{tb}. The other four CKM matrix elements, V_{ud}, V_{us}, V_{cd} and V_{cs} are discussed in [13,3], from where we quote the current best measurements.

- V_{ud} is obtained from the nuclear β-decay of the (super-allowed) $O^+ \to O^+$ transitions [17,18], yielding: $V_{ud} = 0.97377(27)$.
- V_{us} is obtained from the $K_{\ell 3}$ decays and chiral perturbation theory, yielding $V_{us} = 0.22535 \pm 0.00116$. In getting this, the Leutwyler-Roos value for the form factor $f_+(0) = 0.961 \pm 0.008$ [19] has been used. This is to be contrasted with the value, $V_{us} = \lambda = 0.2265(7)$, obtained from the unitarity condition on the first row of the CKM matrix and the input value for V_{ud} given above. They are in good agreement with each other, providing a precision test of the unitarity of the first row of $V_{\rm CKM}$.
- V_{cs} is obtained from the decay $D \to Ke\nu_e$ and the Lattice-QCD result for the $D \to K$ form factor [20], yielding $|V_{cs}| = 0.996 \pm 0.008 \pm 0.015 \pm 0.104$, where the first two errors are experimental and the third (dominant) error is from theory.
- Despite impressive progress in the measurement of the Cabibbo-suppressed D-meson decays and Lattice-QCD, the determination of V_{cd} is still dominated by the neutrino-nucleon production of the charm quark, yielding: $|V_{cd}| = 0.230 \pm 0.011$ [13]. Within errors, $|V_{cd}| = |V_{us}|$, as anticipated from eq. (1).

2.1. *Current determinations of $|V_{cb}|$ and $|V_{ub}|$*

Determinations of $|V_{cb}|$ are based on the semileptonic decay $b \to c\ell\nu_\ell$. This transition can be measured either inclusively through the process $B \to X_c\ell\nu_\ell$, where X_c is a hadronic state with a net c-quantum number, or exclusively, such as the decays $B \to (D, D^*)\ell\nu_\ell$. In either case, intimate knowledge of QCD is required to go from the partonic process to the hadronic states. The fact that $m_b \gg \Lambda_{\rm QCD}$ has led to novel applications of QCD in which heavy quark expansion (HQE) plays a central role [21] and the underlying theory is termed as HQET.

2.1.1. $|V_{cb}|$ *from the decays* $B \to X_c\, \ell\, \nu_\ell$

Concentrating first on the inclusive decays, the semileptonic decay rate can be calculated as a power series

$$\Gamma = \Gamma_0 + \frac{1}{m_b}\Gamma_1 + \frac{1}{m_b^2}\Gamma_2 + \frac{1}{m_b^3}\Gamma_3 + \ldots, \tag{3}$$

where each Γ_i is a perturbation series in $\alpha_s(m_b)$, the QCD coupling constant at the scale m_b. Here Γ_0 is the decay width of a free b-quark, which gives the parton-model result. The coefficient of the leading power correction Γ_1 is absent [22], and the effect of the $1/m_b^2$ correction is collected in Γ_2, which can be expressed in terms of two non-perturbative parameters called λ_1 - kinetic energy of the b-quark - and λ_2 - its chromomagnetic moment. These quantities, also called μ_π^2 and μ_G^2, respectively, in the literature, are defined in terms of the following matrix elements [23–26]

$$\begin{aligned} 2M_B\lambda_1 &\equiv \langle B(v)|\bar{Q}_v(iD)^2Q_v|B(v)\rangle, \\ 6M_B\lambda_2 &\equiv \langle B(v)|\bar{Q}_v\sigma_{\mu\nu}[iD^\mu, iD^\nu]Q_v|B(v)\rangle \ , \end{aligned} \tag{4}$$

where D_μ is the covariant derivative and heavy quark fields are characterized by the 4-velocity, v. At $\mathcal{O}(\Lambda_{\rm QCD}^3/m_b^3)$, six new matrix elements enter in Γ_3, usually denoted by $\rho_{1,2}$ and $\mathcal{T}_{1,2,3,4}$.

Data have been analyzed in the theoretical accuracy in which corrections up to $\mathcal{O}(\alpha_s^2\beta_0)$, $\mathcal{O}(\alpha_s\Lambda_{\rm QCD}/m_b)$ and $\mathcal{O}(\Lambda_{\rm QCD}^3/m_b^3)$ are taken into account [27,28], with β_0 being the lowest order coefficient in the expansion of the QCD β-function. In addition to the other parameters, a quark mass scheme has to be specified. Bauer et al. [27] have carried out a comprehensive study of the scheme dependence using five quark mass schemes: $1S$, PS, $\overline{\rm MS}$, kinematic, and the pole mass.

To extract the value of $|V_{cb}|$ and other fit parameters, three different distributions, namely the charged lepton energy spectrum and the hadronic invariant mass spectrum in $B \to X_c\ell\bar{\nu}_\ell$, and the photon energy spectrum in $B \to X_s\gamma$ have been studied. Theoretical analyses are carried out in terms of the moments. Defining the integral $R_n(E_{\rm cut}, \mu) \equiv \int_{E_{\rm cut}} dE_\ell(E_\ell - \mu)^n d\Gamma/dE_\ell$, where $E_{\rm cut}$ is a lower cut on the charged lepton energy, moments of the lepton energy spectrum are given by $\langle E_\ell^n\rangle = R_n(E_{\rm cut}, 0)/R_0(E_{\rm cut}, 0)$. For the $B \to X_c\ell\bar{\nu}_\ell$ hadronic invariant mass spectrum, the moments are defined likewise with the cutoff $E_{\rm cut}$. Analyses of the data along these lines have been presented by a large number of experiments. For a summary and references to the original literature, see HFAG [7].

The global fit of the world data in the so-called $1S$-scheme for the b-quark mass undertaken by Bauer et al. [27] leads to the following fit values for $|V_{cb}|$ and m_b^{1S}:

$$\begin{aligned} |V_{cb}| &= (41.78 \pm 0.30 \pm 0.08_{\tau_B}) \times 10^{-3}, \\ m_b^{1S} &= (4.701 \pm 0.030)\,\text{GeV}. \end{aligned} \tag{5}$$

The BABAR collaboration have studied the dependence of the lepton and hadron moments on the cutoff $E_{\rm cut}$ and compared their measurements with the theoretical calculation by Gambino and Uraltsev [28] using the so-called kinematic scheme for the b-quark mass $m_b^{\rm kin}(\mu)$, renormalized at the scale $\mu = 1$ GeV. Agreement between experiment and theory allows to determine the fit parameters in this scheme with the results [7]:

$$\begin{aligned} |V_{cb}| &= (41.91 \pm 0.19_{exp} \pm 0.28_{HQE} \pm 0.59_{\Gamma_{\rm sl}}) \times 10^{-3}, \\ m_b\ (1\text{GeV}) &= \ (4.613 \pm 0.022_{exp} \pm 0.027_{HQE})\ \text{GeV}, \\ m_c\ (1\text{GeV}) &= \ (1.187 \pm 0.033_{exp} \pm 0.040_{HQE})\ \text{GeV}\ . \end{aligned} \tag{6}$$

The two analyses (5) and (6) are in excellent agreement with each other. The achieved accuracy $\delta|V_{cb}|/|V_{cb}| \simeq 2\%$ is impressive, and the precision on m_b is also remarkable, $\delta m_b/m_b = O(10^{-3})$, with a similar precision obtained on the mass difference $m_b - m_c$.

2.2. $|V_{cb}|$ *from* $B \to (D, D^*)\ell\nu_\ell$ *decays*

The classic application of HQET in heavy $\to$ heavy decays is in the decay $B \to D^*\ell\nu_\ell$. The differential distribution in the variable $\omega(= v_B.v_{D^*})$, where $v_B(v_{D^*})$ is the four-velocity of the $B(D^*)$-meson, is given by

$$\frac{d\Gamma}{d\omega} = \frac{G_F^2}{4\pi^3} |V_{cb}|^2 m_{D^*}^3 (m_B - m_{D^*})^2 \left(\omega^2 - 1\right)^{1/2} \mathcal{G}(\omega) \left|\mathcal{F}(\omega)\right|^2 ,$$

where $\mathcal{G}(\omega)$ is a phase space factor with $\mathcal{G}(1) = 1$, and $\mathcal{F}(\omega)$ is the Isgur–Wise (IW) function [29] with the normalization at the symmetry point $\mathcal{F}(1) = 1$. Leading $\Lambda_{\rm QCD}/m_b$ corrections in $\mathcal{F}(1)$ are absent due to Luke's theorem [30]. Theoretical issues are the precise determination of the second order power correction to $\mathcal{F}(\omega = 1)$, the slope ρ^2 and the curvature c of the IW-function:

$$\mathcal{F}(\omega) = \mathcal{F}(1)\ \left[1 - \rho^2\,(\omega - 1) + c\,(\omega - 1)^2 + ...\right].$$

Bounds on ρ^2 have been obtained by Bjorken [31] and Uraltsev [32], which can be combined to yield $\rho^2 > 3/4$. Likewise, bounds on the second (and higher) derivatives of the IW-function have been worked out by the Orsay group [33], yielding $c > 15/32$. Apparently, the data sets used in this analysis differ significantly from experiment to experiment, resulting in considerable dispersion in the values of $\mathcal{F}(1)|V_{cb}|$ and ρ^2 and hence in a large χ^2 of the combined fit, summarized by HFAG [7]:

$$\begin{aligned} \mathcal{F}(1)|V_{cb}| &= (35.89 \pm 0.56) \times 10^{-3}\ , \\ \rho^2 &= 1.23 \pm 0.05 \quad (\chi^2 = 37.8/17; \text{CL} = 0.026). \end{aligned} \tag{7}$$

To convert this into a value of $|V_{cb}|$, we need to know $\mathcal{F}(1)$. In terms of the perturbative (QED and QCD) and non-perturbative (leading δ_{1/m^2} and sub-leading δ_{1/m^3}) corrections, $\mathcal{F}(1)$ can be expressed as follows:

$$\mathcal{F}(1) = \eta_A \left[1 + \delta_{1/m^2} + \delta_{1/m^3}\right] , \tag{8}$$

where η_A is the perturbative renormalization of the IW-function, known in the meanwhile to three loops [34]. One- and two-loop corrections yield $\eta_A \simeq 0.933$ and the $O(\alpha_s^3)$ contribution amounts to $\eta_A^{(3)} = -0.005$. A Lattice-QCD calculation in the quenched approximation yields [35] $\mathcal{F}(1) = 0.919^{+0.030}_{-0.035}$, which is now being reevaluated with dynamical quarks. Taking into account this theoretical input, the value quoted at the Lepton-Photon-2007 Symposium is [3]:

$$|V_{cb}|_{B\to D^*\ell\nu_\ell} = (39.1 \pm 0.65_{\text{exp}} \pm 1.4_{\text{theo}}) \times 10^{-3}. \tag{9}$$

The resulting value of $|V_{cb}|$ is in excellent agreement with the ones given in (5) and (6) obtained from the inclusive decays.

2.3. $|V_{ub}|$ *from the decays* $B \to X_u\ell\nu_\ell$

HQET techniques allow to calculate the inclusive decay rate $B \to X_u\ell\nu_\ell$ rather accurately. However, the experimental problem in measuring this transition lies in the huge background from the dominant decays $B \to X_c\ell\nu_\ell$ which can be brought under control only through severe cuts on the kinematics. For example, these cuts are imposed on the lepton energy. demanding $E_\ell > (m_B^2 - m_D^2)/2m_B$, and/or the momentum transfer to the lepton pair q^2 restricting it below a threshold value $q^2 < q^2_{\text{max}}$, and/or the hadron mass recoiling against the leptons, which is required to satisfy $m_X < m_D$. With these cuts, the phase space of the decay $B \to X_u\ell\nu_\ell$ is greatly reduced. A bigger problem is encountered in the end-point region (also called the shape function region), where the leading power correction is no longer $1/m_b^2$ but rather $1/m_b\Lambda_{\text{QCD}}$, slowing the convergence of the expansion. Moreover, in the region of energetic leptons with low invariant mass hadronic states, the differential rate is sensitive to the details of the shape function $f(k_+)$ [36], where $k_+ = k_0 + k_3$ with $k^\mu \sim O(\Lambda_{\text{QCD}})$.

The need to know $f(k_+)$ can be circumvented to a large extent by doing a combined analysis of the data on $B \to X_u\ell\nu_\ell$ and $B \to X_s\gamma$. Using the operator product expansion (OPE) to calculate the photon energy spectrum in the inclusive decay $B \to X_s\gamma$, the leading terms in the spectrum (neglecting the bremsstrahlung corrections) can be re-summed into a shape function [37]:

$$\frac{d\Gamma_s}{dx} = \frac{G_F^2 \alpha m_b^5}{32\pi^4} |V_{ts}V_{tb}^*|^2 \, |C_7^{\text{eff}}|^2 \, f(1-x) , \tag{10}$$

where $x = \frac{2E_\gamma}{m_b}$, and $C_7^{\rm eff}$ is an effective Wilson coefficient, characterizing the strength of the electromagnetic dipole operator. In the leading order, E_ℓ- and M_{X_u}-spectra in $B \to X_u \ell \nu_\ell$ are also governed by $f(x)$. Thus, $f(x)$ can be measured in $B \to X_s \gamma$ and used in the analysis of data in $B \to X_u \ell \nu_\ell$.

Following this argument, a useful relation emerges [38–40]

$$|\frac{V_{ub}}{V_{tb}V_{ts}^*}| = \left(\frac{3\alpha}{\pi} |C_7^{\rm eff}|^2 \frac{\Gamma_u(E_c)}{\Gamma_s(E_c)} \right)^{\frac{1}{2}} (1 + \delta(E_c)) \; , \tag{11}$$

where

$$\begin{aligned} \Gamma_u(E_c) &\equiv \int_{E_c}^{m_B/2} dE_\ell \frac{d\Gamma_u}{dE_\ell} \; , \\ \Gamma_s(E_c) &\equiv \frac{2}{m_b} \int_{E_c}^{m_B/2} dE_\gamma (E_\gamma - E_c) \frac{d\Gamma_s}{dE_\gamma} \; , \end{aligned} \tag{12}$$

and $\delta(E_c)$ incorporates the sub-leading terms in $\mathcal{O}(\Lambda_{\rm QCD}/m_b)$, which can only be modeled at present. In addition, there are perturbative corrections to the spectra and in the relation (11) [37,41,38].

The other strategy is to extend the measurements of the inclusive decay $B \to X_u \ell^+ \nu_\ell$ to the kinematic regions which are reachable by the decay $B \to X_c \ell^+ \nu_\ell$, thus enlarging the region where the light-cone momentum component satisfies $P_+ \equiv E_X - |\vec{P}_X| \gg \Lambda_{\rm QCD}$, obviating the need to know the shape function. Both these methods have been used in the determination of $|V_{ub}|$, summarized below.

• Determination of $|V_{ub}|$ using the combined cuts on the variables m_X and q^2 following the suggestion by Bauer, Ligeti and Luke (BLL) [42]. Taking the HQE parameter input from the analysis of the decay $B \to X_c \ell \nu_\ell$ and $B \to X_s \gamma$, HFAG [7] quotes an average value $|V_{ub}|(\text{BLL}) = (4.83 \pm 0.24 \pm 0.37) \times 10^{-3}$, using the b-quark mass $m_b(1\text{S}) = (4.70 \pm 0.03)$ GeV in the so-called $1S$-scheme.

• Determination of $|V_{ub}|$ from a fully differential decay rate for $B \to X_u \ell \nu_\ell$ based on the soft collinear effective theory (SCET) techniques [43–46], hereinafter called the Bosch-Lange-Neubert-Paz (BLNP) approach [47]. The three independent kinematic variables are chosen to be: $P_\ell = M_B - 2E_\ell$, $P_- = E_X + |\vec{P}_X|$, and $P_+ = E_X - |\vec{P}_X|$, where $P_\pm$ are the light-cone components of the hadronic final-state momentum along the jet direction, E_X is the jet energy, $\vec{P}_X$ is the jet momentum, and E_ℓ is the charged-lepton energy. In terms of these variables, the triple differential distribution is:

$$\frac{d^3\Gamma}{dP_+dP_-dP_\ell} = \frac{G_F^2|V_{ub}|^2}{16\pi^2}(M_B - P_+)\{(P_- + P_+)(M_B - P_- + P_\ell - P_+)F_1 +(M_B - P_-)(P_- - P_+)F_2 + (P_- - P_\ell)(P_\ell - P_+)F_3\}. \quad (13)$$

The "structure functions" F_i can be expressed as product of the hard (perturbatively calculable) coefficient and a jet function, which are convoluted with the soft light-cone distribution functions, the shape functions of the B meson. SCET allows to separate the two scales here, namely $\mu_h \sim m_b$ and $\mu_i \sim \sqrt{m_b\Lambda_{\rm QCD}}$ and enables to sum large logarithms involving the two scales μ_h and μ_i. The dependence on the subleading shape functions is studied by taking several models. Fixing the HQE parameters (in the so-called shape function scheme) to the values $m_b(\text{SF}) = (4.63 \pm 0.06)$ GeV, $\mu_\pi^2(\text{SF}) = (0.18 \pm 0.06)$ GeV^2, and the exponential form for the shape function, HFAG [7] quotes $|V_{ub}| = (4.31 \pm 0.17 \pm 0.35) \times 10^{-3}$ in the BLNP approach.

• Determination of $|V_{ub}|$ using the so-called Dressed Gluon Exponentiation (DGE) advocated by Andersen and Gardi [48]. The basic assumption of this approach is that properly defined quark distribution in an on-shell heavy quark provides a good approximation to the distribution in the meson. The problematic small hadronic mass M_{X_u} region in the decay $B \to X_u\ell^+\nu_\ell$ is characterized by a large hierarchy in the ratio $p_j^+/p_j^- \ll 1$ involving the partonic light-cone coordinates $p_j^+ \leq p_j^- \leq m_b$. Defining the moments n with respect to the powers of $1 - p_j^+/p_j^-$, the region of small p_j^+ is probed by high moments $n \to \infty$, giving rise to the Sudakov logarithms $\ln n$. Infrared sensitivity appears in the moment-space Sudakov exponents through infrared renormalons, leading to the divergence of the (higher order perturbative) series in the Sudakov exponent. However, the leading infrared renormalon ambiguity cancels exactly against the pole-mass renormalon ambiguity, which enters as a kinematic factor in the differential decay width, making the on-shell perturbative calculation directly applicable for the phenomenology. This approach, applied to the data yields [7] $|V_{ub}| = (4.34 \pm 0.16 \pm 0.25) \times 10^{-3}$ for the input value $\overline{m_b}(\bar{m}_b) = (4.20 \pm 0.07)$ GeV.

Thus, the three theoretical approaches used in the determination of $|V_{ub}|$ from the inclusive decays $B \to X_u\ell\nu_\ell$ give very consistent values.

2.4. $|V_{ub}|$ *from exclusive decays*

$|V_{ub}|$ has also been determined from the exclusive decays $B \to \pi \ell \nu_\ell$. Theoretical accuracy is limited by the imprecise knowledge of the form factors. A number of theoretical techniques has been used to determine them. These include, among others, Light-cone QCD sum rules [49], Quenched- and Unquenched-Lattice QCD simulations [50–52]. HFAG quotes the following extracted values of $|V_{ub}|$ using the full q^2-range for the form factors (in units of 10^{-3}) [7]:
$|V_{ub}| = 3.43 \pm 0.10^{+0.67}_{-0.42}$ (LCSR: Ball-Zwicky[49]);
$|V_{ub}| = 3.17 \pm 0.10^{+0.77}_{-0.48}$ (Lattice: HPQCD[50]);
$|V_{ub}| = 3.82 \pm 0.12^{+0.88}_{-0.52}$ (Lattice: FNAL[51]);
$|V_{ub}| = 3.61 \pm 0.11^{+1.11}_{-0.57}$ (Lattice: APE[52]).

These values of $|V_{ub}|$ from the exclusive decays (typically 3.5×10^{-3}) are smaller than the ones extracted from the inclusive measurements (typically 4.3×10^{-3}). This mismatch is a mild irritant, waiting for its resolution through progress in Lattice-QCD and more B-factory data.

To reduce the form-factor related uncertainties in extracting $|V_{ub}|$ from exclusive decays $B \to (\pi, \rho)\ell\nu_\ell$, input from the rare B-decays $B \to (K, K^*)\ell^+\ell^-$ and HQET may be helpful. A proposal along these lines is the so-called Grinstein's double ratio which would determine $|V_{ub}|/|V_{tb}V_{ts}^*|$ from the end-point region of exclusive rare B-meson decays [53]. To carry out this program one has to measure four distributions in the decays $B \to \rho\ell\nu_\ell$, $B \to K^*\ell^+\ell^-$, and $D \to (\rho, K^*)\ell\nu_\ell$. With the help of this data and HQET, the ratio of the CKM factors $|V_{ub}|/|V_{tb}V_{ts}^*|$ can be determined through the double ratio

$$\frac{\Gamma(\bar{B} \to \rho\ell\nu)}{\Gamma(\bar{B} \to K^*\ell^+\ell^-)} \, \frac{\Gamma(D \to K^*\ell\nu)}{\Gamma(D \to \rho\ell\nu)} . \tag{14}$$

At the B factories, one expects enough data on these decays to allow a 10% determination of $|V_{ub}|$ from exclusive decays.

3. Status of the Third Row of $V_{\rm CKM}$

FCNC transitions $b \to s$ and $b \to d$ (as well as $s \to d$) give information on the third row of the CKM matrix $V_{\rm CKM}$ and allow to search for physics beyond the SM. In the SM these transitions are generally dominated by the (virtual) top quark contributions giving rise to the dependence on the matrix elements $|V_{tb}^*V_{ts}|$ (for $b \to s$ transitions) and $|V_{tb}^*V_{td}|$ (for $b \to d$ transitions). Of these, only the matrix element $|V_{tb}|$ has been measured by a tree process $t \to Wb$ at the Tevatron through the ratio

$$R_{tb} \equiv \frac{\mathcal{B}(t \to Wb)}{\mathcal{B}(t \to Wq)} = \frac{|V_{tb}|^2}{|V_{td}|^2 + |V_{ts}|^2 + |V_{tb}|^2} \,. \tag{15}$$

The current measurements yield [54]: $R_{tb} = 1.12^{+0.21}_{-0.19}(\text{stat})^{+0.17}_{-0.13}(syst)$, yielding $R_{tb} > 0.61$ (at 95% C.L.), which in turn gives $|V_{tb}| > 0.78$ at 95% C.L..

Precision on the direct measurement of $|V_{tb}|$ is rather modest (unitarity gives $|V_{tb}| \simeq 0.9992$.), which will be greatly improved, in particular, at a Linear Collider [55], but also at the LHC. The corresponding measurements of $|V_{ts}|$ and $|V_{td}|$ from the tree processes are not on the cards. These matrix elements are determined by (loop) induced processes discussed below.

3.1. *Status of* $|V_{td}|$

The current best measurement of $|V_{td}|$ comes from ΔM_{B_d}, the mass difference between the two mass eigenstates of the B^0_d - $\overline{B^0_d}$ complex. This has been measured in a number of experiments and is known to an accuracy of $\sim 1\%$; the current world average is [7] $\Delta M_{B_d} = 0.507 \pm 0.005$ (ps)$^{-1}$.

In the SM, ΔM_{B_d} and its counterpart ΔM_{B_s}, the mass difference in the B^0_s - $\overline{B^0_s}$ system, are calculated by box diagrams, dominated by the Wt loop. Since $(M_W, m_t) \gg m_b$, ΔM_{B_d} is governed by the short-distance physics. The expression for ΔM_{B_d} taking into account the perturbative-QCD corrections reads as follows [56]

$$\Delta M_{B_d} = \frac{G_F^2}{6\pi^2}\, \hat{\eta}_B \, |V_{td} V_{tb}^*|^2 \, M_{B_d} \, (f_{B_d}^2 \hat{B}_{B_d}) \, M_W^2 \, S_0(x_t) \,. \tag{16}$$

The quantity $\hat{\eta}_B$ is the next-to-leading log (NLL) perturbative QCD renormalization of the matrix element of the $(|\Delta B| = 2, \Delta Q = 0)$ four-quark operator, whose value is $\hat{\eta}_B = 0.55 \pm 0.01$ [57]; $x_t = m_t^2/M_W^2$ and $S_0(x_t) = x_t f_2(x_t)$ is an Inami-Lim function [58], with

$$f_2(x) = \frac{1}{4} + \frac{9}{4}\frac{1}{(1-x)} - \frac{3}{2}\frac{1}{(1-x)^2} - \frac{3}{2}\frac{x^2 \ln x}{(1-x)^3} \,. \tag{17}$$

The quantity $f_{B_d}^2 \hat{B}_{B_d}$ enters through the hadronic matrix element of the four-quark box operator, defined as:

$$\langle \bar{B}_q^0 | (\bar{b}\gamma_\mu (1-\gamma_5) q)^2 | B_q^0 \rangle \equiv \frac{8}{3} f_{B_q}^2 B_{B_q} M_{B_q}^2 \,, \tag{18}$$

with $B_q = B_d$ or B_s. With ΔM_{B_d} and $\hat{\eta}_B$ known to a high accuracy, and the current value of the top quark mass, defined in the $\overline{\text{MS}}$ scheme, $\bar{m}_t(m_t) = (162.3 \pm 2.2)$ GeV, leading to $S_0(x_t) = 2.29(5)$, i.e., $\delta S_0(x_t)/S_0(x_t) \simeq 2.5\%$, the combined error from these sources is a few percent. This is completely negligible in comparison with the current theoretical uncertainty on the matrix element $f_{B_d}\sqrt{\hat{B}_{B_d}}$. For example, $O(\alpha_s)$-improved calculations in the QCD sum rule approach yield $f_{B_d} = (210\pm19)$ MeV [59] and $f_{B_d} = (206\pm20)$ MeV [60], whereas $\overline{B}_{B_d}$ in the $\overline{\rm MS}$ scheme in this approach is estimated as [61] $\overline{B}_{B_d} = 1$ to within 10%, yielding for the renormalization group invariant quantity $\hat{B}_{B_d} \simeq 1.46$, and an accuracy of about $\pm15\%$ on $f_{B_d}\sqrt{\hat{B}_{B_d}}$. Lattice calculations for $f_{B_d}\sqrt{\hat{B}_{B_d}}$ are uncertain due to the chiral extrapolation. Taking this into account, the current unquenched lattice QCD calculations from the JLQCD and HPQCD Collaborations yield [62,63]

$$f_{B_d}\sqrt{\hat{B}_{B_d}} = (244 \pm 11 \pm 24)\ \text{MeV}\,, \tag{19}$$

where the first error is statistical, and the second combines the uncertainty from chiral extrapolation and other systematic errors. Using $|V_{tb}| = 1$, this yields [64]

$$|V_{td}| = (7.4 \pm 0.8) \times 10^{-3}\,. \tag{20}$$

3.2. *Present status of* $|V_{ts}|$

The quantity which currently provides the best determination of $|V_{ts}|$ is ΔM_{B_s}, measured by D0 and CDF, with the CDF measurement being more precise $\Delta M_{B_s} = (17.77\pm0.10\pm0.07)\ (\text{ps})^{-1}$ [11]. The expression for ΔM_{B_s} in the SM can be obtained from the one for ΔM_{B_d} (16) by the replacements: $M_{B_d} \to M_{B_s}$, $(f_{B_d}^2 \hat{B}_{B_d}) \to (f_{B_s}^2 \hat{B}_{B_s})$, and, most importantly, $V_{td} \to V_{ts}$. Using the recent Lattice-QCD result $f_{B_s}\sqrt{\hat{B}_{B_s}} = (281 \pm 21)$ MeV from the HPQCD collaboration [65] allows to determine $|V_{ts}|$ from ΔM_{B_s} with a precision of about 10%, comparable to the precision on $|V_{td}|$ in eq. (20). This is not competitive with the indirect estimate of this matrix element from the CKM unitarity, which yields $V_{ts} \simeq -V_{cb} = -4.1(1) \times 10^{-2}$.

The compatibility of the SM with the measured value of ΔM_{B_s} is usually tested by taking the value of V_{ts} from the unitarity fits, the measured value of m_t, and the Lattice-QCD value for $f_{B_s}\sqrt{\hat{B}_{B_s}}$. Following this reasoning, the Lattice-HPQCD collaboration estimates $\Delta M_{B_s}(\text{SM}) = (20.3\pm3.0\pm0.8)$ $(\text{ps})^{-1}$ [65]. The corresponding estimate from the UTfit is $\Delta M_{B_s}(\text{SM}) = (20.9 \pm 2.6)\ (\text{ps})^{-1}$ [66] and the CKMfitter yields $\Delta M_{B_s}(\text{SM}) = (21.7^{+5.9}_{-4.2})$ $(\text{ps})^{-1}$ [67]. Thus, the experimental measurement of ΔM_{B_s} is typically about 1σ below the SM estimates, with $\Delta M_s(\text{expt})/\Delta M_s(\text{SM}) = 0.88 \pm 0.13$ [65], 0.85 ± 0.10 [66] and 0.88 ± 0.20 [67].

Possible implications of the CDF measurement of ΔM_{B_s} have been studied in several papers [68–71,66]. Perhaps, it is to the point to mention here that a value of $\Delta M_{B_s} \simeq 18\ (\mathrm{ps})^{-1}$ was also hinted by the LEP data with the central value being $17.7\ (\mathrm{ps})^{-1}$. Anticipating this, the consequences of an eventual measurement of ΔM_{B_s} around this value were worked out some eight years ago [72] for the parameters of the SM and the minimal flavor-violating supersymmetry [73]. It was emphasized that a measurement of ΔM_{B_s} around this value would confirm the SM. The CDF measurement of ΔM_{B_s} is very precise, and our knowledge of the CKM parameters and the non-perturbative quantities has in the meanwhile also improved, as discussed in this report. However, the bottom line remains the same, namely that the SM has passed this crucial test comfortably.

The ratio of the mass differences $\Delta M_{B_d}/\Delta M_{B_s}$, now measured very precisely, can be used to constrain the CKM ratio $|V_{td}|/|V_{ts}|$ using the SM relation [74]:

$$\frac{\Delta M_{B_s}}{\Delta M_{B_d}} = \xi \frac{M_{B_s}}{M_{B_d}} \frac{|V_{tb}^* V_{ts}|^2}{|V_{tb}^* V_{td}|^2}, \tag{21}$$

where $\xi \equiv f_{B_s}\sqrt{\hat{B}_{B_s}}/f_{B_d}\sqrt{\hat{B}_{B_d}}$. Theoretical uncertainty in ξ in the Lattice QCD approach is arguably smaller compared to the one in $f_{B_s}\sqrt{\hat{B}_{B_s}}$, as in the SU(3) limit $\xi = 1$, and the uncertainty is actually in the SU(3)-breaking corrections. Current estimate in the unquenched lattice calculations of ξ is [64] $\xi = 1.21^{+0.047}_{-0.035}$, which yields [11]

$$|V_{td}/V_{ts}| = 0.2060 \pm 0.0007(\mathrm{exp})^{+0.008}_{-0.006}(\mathrm{th}). \tag{22}$$

This is by far the best measurement of this CKM ratio, and it provides a non-trivial constraint on the allowed profile of the unitarity triangle. Combining eqs. (22) and (20) yields $|V_{ts}| = (36 \pm 4) \times 10^{-3}$ with the error dominated by theory. This completes our review of the CKM matrix elements V_{ij}.

4. Radiative and Semileptonic Rare B Decays

Two inclusive rare B-decays of current experimental interest are $B \to X_s\gamma$ and $B \to X_s l^+l^-$, where X_s is any charmless hadronic state with the strangeness quantum number $s = 1$. They probe the SM in the electroweak $b \to s$ penguin sector. The CKM-suppressed decays $B \to X_d\gamma$ and $B \to X_d l^+l^-$ are difficult to measure due to low rates and formidable backgrounds. Instead, the search for $B \to X_d\gamma$ decay has been carried out in the exclusive decay modes $B \to (\rho, \omega)\gamma$. Combined with the decay $B \to K^*\gamma$, these decays provide constraints on the CKM parameters. The CKM-suppressed decays $B \to (\pi, \rho, \omega)\ell^+\ell^-$ have not yet been measured. We review some of these rare B-decays in the context of the SM.

4.1. $B \to X_s\gamma$: *SM vs. Experiments*

The effective Lagrangian for the decays $B \to X_s\gamma$ obtained by integrating out the top quark and the heavy electroweak bosons reads as follows in the SM:

$$\mathcal{L}_{\rm eff} = \frac{4G_F}{\sqrt{2}} V_{ts}^* V_{tb} \sum_{i=1}^{8} C_i(\mu) Q_i \, . \tag{23}$$

In writing this, unitarity of the CKM matrix has been used and the term proportional to the small matrix elements $V_{us}^* V_{ub}$ has been ignored. The complete list of operators and their Wilson coefficients in the NNLO approximation evaluated at the scale $\mu = m_b$ can be seen elsewhere [75].

The dominant four-quark operators Q_1 and Q_2 are

$$Q_1 = (\bar{s}\, c)_{V-A}\, (\bar{c}\, b)_{V-A}, \qquad Q_2 = (\bar{s}_i c_j)_{V-A}\, (\bar{c}_j b_i)_{V-A}, \tag{24}$$

and the electromagnetic and chromomagnetic penguin operators Q_7 and Q_8 are

$$Q_7 = -\frac{e\, \overline{m}_b(\mu)}{8\pi^2} \, (\bar{s}\, \sigma^{\mu\nu} \, [1+\gamma_5]\, b) F_{\mu\nu} \, , \quad Q_8 = -\frac{g\, \overline{m}_b(\mu)}{8\pi^2} \, (\bar{s}\, \sigma^{\mu\nu} \, [1+\gamma_5]\, T^a \, b) G^a_{\mu\nu}. \tag{25}$$

The factor $\overline{m_b}(\mu)$ is the $\overline{\rm MS}$ mass of the b quark. QCD-improved calculations in the effective theory require three steps (for a review see [76]):
(i) Matching $C_i(\mu_0)$ $(\mu_0 \sim M_W, m_t)$: They have been calculated up to three loops [77,78]. The three-loop matching is found to have less than 2% effect on $\mathcal{B}(B \to X_s\gamma)$ [78].
(ii) Operator mixing: This involves calculation of the anomalous dimension matrix, which is expanded in $\alpha_s(\mu)$. The anomalous dimensions up to $\alpha_s^2(\mu)$ are known [79] since a decade, and the $\alpha_s^3(\mu)$ calculations have been completed recently in a series of papers [80–82].
(iii) Matrix elements $\langle O_i \rangle(\mu_b)$ $(\mu_b \sim m_b)$: The first two terms in the expansion in $\alpha_s(\mu_b)$ are known since long [83]. Exact results to $\mathcal{O}(\alpha_s^2)$ were obtained for Q_7 in [84,85] and for Q_8 in [86]. For Q_1 the virtual corrections at $\mathcal{O}(\alpha_s)$ were calculated in [83,87,88], but those at $\mathcal{O}(\alpha_s^2)$ are known only in the large-β_0 limit [89]. A calculation that goes beyond this approximation by

employing an interpolation in the charm quark mass m_c was reported in [90], and has been used in estimating the NNLO branching fraction for the inclusive decay $B \to X_s\gamma$ [91]. Finally, one has to add the Bremsstrahlung contribution $b \to s\gamma g$ to get the complete decay rate, which in $O(\alpha_s)$ was done in [92] and in $O(\alpha_s^2)$ in [93].

In the $\overline{\text{MS}}$ scheme, the NNLO branching ratio for $E_\gamma > 1.6$ GeV is calculated as [91]:

$$\mathcal{B}(B \to X_s\gamma)_{\text{SM}} = (3.15 \pm 0.23) \times 10^{-4} . \tag{26}$$

This amounts to a theoretical precision of about 8%, comparable to the current experimental precision [7]

$$\mathcal{B}(B \to X_s\gamma)_{\text{Expt.}} = (3.55 \pm 0.24^{+0.09}_{-0.10} \pm 0.03) \times 10^{-4} . \tag{27}$$

A comparison of the two shows that the SM estimate is in agreement with data though the SM central value lies below the experiment by about 1σ. This allows for speculations about a beyond-the-SM contribution interfering constructively with the SM amplitude. A case in point is a 2Higgs doublet model (2HDM); the preferred value is $m_{H^+} \simeq 650$ GeV with a 95% C.L. lower bound $m_{H^+} > 295$ GeV. However, more conservatively, the proximity of $\mathcal{B}(B \to X_s\gamma)$ in the SM and experiment puts bounds on the parameters characterizing new physics. This has been worked out, together with other constraints, in the context of supersymmetry [94].

The current (NNLO) theoretical precision on $\mathcal{B}(B \to X_s\gamma)$ has also been investigated in the context of SCET using a multi-scale OPE involving three low energy scales: m_b, $\sqrt{m_b\Delta}$ and $\Delta = m_b - 2E_0$, where E_0 is the lower cut on the photon energy. Large logarithms associated with these scales are summed at NLL order. The sensitivity to the scale Δ introduces additional uncertainties. Taking this into account, Becher and Neubert [95] estimate $\mathcal{B}(B \to X_s\gamma) = (2.98 \pm 0.26) \times 10^{-4}$, which increases the departure of the SM from data to about 1.4σ.

4.2. $B \to X_s\ell^+\ell^-$: *SM vs. Experiments*

To study the decays $B \to X_s\ell^+\ell^-$, one has to extend the operator basis in the effective Lagrangian (23) by adding two semileptonic operators [96]:

$$Q_9 = \frac{e^2}{16\pi^2}(\bar{s}\,\gamma_\mu\, b)_{V-A}\,(\bar{\ell}\,\gamma^\mu\,\ell)\,, \qquad Q_{10} = \frac{e^2}{16\pi^2}(\bar{s}\,\gamma_\mu\, b)_{V-A}\,(\bar{\ell}\,\gamma^\mu\gamma_5\,\ell)\,.$$

The corresponding Wilson coefficients $C_9(\mu)$ and $C_{10}(\mu)$ have the following perturbative expansion:

$$C_9 = \frac{4\pi}{\alpha_s(\mu)} C_9^{(-1)}(\mu) + C_9^{(0)}(\mu) + \frac{\alpha_s(\mu)}{4\pi} C_9^{(1)}(\mu) + ...$$
$$C_{10} = C_{10}^{(0)} + \frac{\alpha_s(M_W)}{4\pi} C_{10}^{(1)} + ... \quad (28)$$

After an expansion in α_s, the term $C_9^{(-1)}(\mu)$ reproduces the dominant part of the electroweak logarithms that originate from photonic penguins with charm quark loops:

$$\frac{4\pi}{\alpha_s(m_b)} C_9^{(-1)}(m_b) = \frac{4}{9} \ln \frac{M_W^2}{m_b^2} + O(\alpha_s) , \quad (29)$$

leading to $\frac{4\pi}{\alpha_s(m_b)} C_9^{(-1)}(m_b) \simeq 2$. With $C_9^{(0)}(m_b) \simeq 2.2$, one needs to calculate in the NNLO accuracy. The NNLO calculation of the decay $B \to X_s l^+ l^-$ corresponds to the NLO calculation of $B \to X_s \gamma$, as far as the number of loops in the diagrams is concerned.

The process $B \to X_s \ell^+ \ell^-$ differs greatly from the radiative decay $B \to X_s \gamma$ as far as non-perturbative contributions are concerned. The largest effect in $B \to X_s \ell^+ \ell^-$ from the intermediate $c\bar{c}$ states comes from the resonances J/ψ, ψ' and ψ'' decaying to $\ell^+ \ell^-$, which can be either modeled, for example, as done by Krüger and Sehgal [97] using dispersion relations and data on $\sigma_(e^+ e^- \to c\bar{c} \to$ hadrons), or else experimental cuts are imposed on q^2 to remove the resonant regions and the short-distance contribution is extrapolated through these cuts. Then, there are factorizable $1/m_c$ and $1/m_b$ power corrections, similar to those in $B \to X_s \gamma$, which can be calculated using the OPE and HQET. As is the case for $B \to X_s \gamma$ and $B \to X_u \ell \nu_\ell$, there are no $1/m_b$ corrections. The $O(1/m_b^2)$ corrections in this framework were calculated first in [98] and corrected in [99]. The $O(1/m_b^3)$ corrections were calculated in [100]. The $1/m_c$ factorizable power corrections were calculated in [101].

Including the leading power corrections in $1/m_b$ and $1/m_c$ and taking into account various parametric uncertainties, the branching ratios for the decays $B \to X_s \ell^+ \ell^-$ in NNLO are [102]:

$$\mathcal{B}(B \to X_s e^+ e^-)_{\rm SM} \simeq \mathcal{B}(B \to X_s \mu^+ \mu^-)_{\rm SM} = (4.2 \pm 0.7) \times 10^{-6} , \quad (30)$$

where a dilepton invariant mass cut, $m_{\ell\ell} > 0.2$ GeV, has been assumed for comparison with data given below. These estimates make use of the NNLO calculation by Asatryan et al. [103], restricted to $\hat{s} \equiv q^2/m_b^2 < 0.25$. The spectrum for $\hat{s} > 0.25$ has been obtained from the NLO calculations using the scale $\mu_b \simeq m_b/2$, as this choice of the scale reduces the NNLO contributions. Subsequent NNLO calculations covered the entire dilepton mass spectrum and are numerically in agreement with this procedure, yielding $\mathcal{B}(B \to X_s\mu^+\mu^-)_{\rm SM} = (4.6 \pm 0.8) \times 10^{-6}$ [104,105] . The difference in the central values in these results and (30) is of parametric origin.

The BABAR and BELLE collaborations have measured the invariant dilepton and hadron mass spectra in $B \to X_s\ell^+\ell^-$. Using the SM-based calculations to extrapolate through the cut-regions, the current averages of the branching ratios are [7]:

$$\begin{aligned} \mathcal{B}(B \to X_s e^+e^-) &= (4.7 \pm 1.3) \times 10^{-6}, \\ \mathcal{B}(B \to X_s \mu^+\mu^-) &= (4.3^{+1.3}_{-1.2}) \times 10^{-6} , \\ \mathcal{B}(B \to X_s \ell^+\ell^-) &= (4.5^{+1.03}_{-1.01}) \times 10^{-6}. \end{aligned} \tag{31}$$

Thus, within the current experimental accuracy, which is typically 25%, data and the SM agree with each other in the $b \to s$ electroweak penguins. The low q^2-region (say, $q^2 < 8$ GeV^2), which allows the most precise comparison with the SM, suffers both from the statistics and a cut on the invariant hadronic mass recoiling against the dilepton. A cut $m_X > 2GeV$ and $m_X > 1,8$ GeV have been used by the BELLE and BABAR collaborations, respectively. The effects of these cuts have been studied in the Fermi-motion model [106], which has been used in the experimental analysis of the data so far. Subsequently, the $B \to X_s\ell^+\ell^-$ rate with an m_X cut in the low-q^2 region has been calculated using the $B \to X_s\gamma$ shape function [107]. This work, whose impact on the analysis of the $B \to X_s\ell^+\ell^-$ data has yet to be studied, reduces some of the theoretical errors in the SM estimates given in eq. (30). In the same vein, it has also been recently argued [108] that the non-perturbative uncertainties in the large-q^2 region ($q^2 \geq 14$ GeV^2) can be significantly reduced by normalizing the partial decay width of $B \to X_s\ell^+\ell^-$ with the corresponding partial width of the decay $B \to X_u\ell\nu_\ell$. With more data from the B factories, these theoretical developments will enable a more precise test of the SM in the $B \to X_s\ell^+\ell^-$ decays.

The measurements (27) and (31) provide valuable constraints on beyond-the-SM physics scenarios. Following the earlier analysis to determine the Wilson coefficients in $b \to s$ transitions [109,102,110], it has been recently argued [111] that data now disfavor solutions in which the coefficient $C_7^{\rm eff}$ is similar in magnitude but opposite in sign to the SM coefficient.

Exclusive decays $B \to (K, K^*)\ell^+\ell^-$ ($\ell^\pm = e^\pm, \mu^\pm$) have also been measured by the BABAR and BELLE collaborations, and the current world averages of the branching ratios are [7]:

$$\begin{aligned}
\mathcal{B}(B \to K\ell^+\ell^-) &= (3.9 \pm 0.6) \times 10^{-7}, \\
\mathcal{B}(B \to K^* e^+ e^-) &= (11.3^{+2.8}_{-2.6}) \times 10^{-7}, \\
\mathcal{B}(B \to K^* \mu^+ \mu^-) &= (10.3^{+2.6}_{-2.3}) \times 10^{-7}, \qquad (32)\\
\mathcal{B}(B \to K^* \ell^+ \ell^-) &= (9.4^{+1.7}_{-1.6}) \times 10^{-7}. \qquad (33)
\end{aligned}$$

They are also in agreement with the SM-based estimates of the same. A calculation based on the light cone QCD sum rules for the form factors [112] yields [102]: $\mathcal{B}(B \to K\ell^+\ell^-) = (3.5 \pm 1.2) \times 10^{-7}$, $\mathcal{B}(B \to K^* e^+ e^-) = (15.8 \pm 4.9) \times 10^{-7}$, and $\mathcal{B}(B \to K^* \mu^+ \mu^-) = (11.9 \pm 3.9) \times 10^{-7}$ with the errors dominated by uncertainties on the form factors. In the future, these errors can be reduced by using the data on $B \to (\pi, \rho)\ell\nu_\ell$ to determine the $B \to (\pi, \rho)$ form factors. This information can be combined with estimates of the SU(3)-symmetry breaking to determine the $B \to (K, K^*)$ form factors, enabling to predict the FCNC decay rates and spectra more precisely. For the low invariant mass of the dileptons, say $q^2 < 8$ GeV^2, the SCET framework can be employed to reduce the number of form factors and improve the perturbative aspects of these decays.

The Forward-Backward (FB) asymmetry in the decay $B \to X_s\ell^+\ell^-$, defined as [112]

$$\bar{\mathcal{A}}_{\mathrm{FB}}(q^2) = \frac{1}{d\mathcal{B}(B \to X_s\ell^+\ell^-)/dq^2} \int_{-1}^{1} d\cos\theta_\ell \, \frac{d^2\mathcal{B}(B \to X_s\ell^+\ell^-)}{dq^2 \, d\cos\theta_\ell} \mathrm{sgn}(\cos\theta_\ell) ,$$

provides additional constraints on the Wilson coefficients. In particular, the location of the zero-point of this asymmetry (called below q_0^2) is a precision tests of the SM. In NNLO, one has the following predictions for the inclusive decays $B \to X_s\ell^+\ell^-$: $q_0^2 = (3.90 \pm 0.25)$ GeV^2 [$(3.76 \pm 0.22_{\text{theory}} \pm 0.24_{m_b})$ GeV^2], obtained by Ghinculov et al. [113] [Asatrian et al. [114]].

In the SM (and its extensions in which the operator basis remains unchanged), the FB-asymmetry in $B \to K\ell^+\ell^-$ is zero and in $B \to K^*\ell^+\ell^-$ it depends on the decay form factors. Model-dependent studies yield small form factor-related uncertainties in the zero-point of the asymmetry $\hat{s}_0 = q_0^2/m_B^2$ [115]. HQET provides a symmetry argument why the uncertainty in $\hat{s}_0$ can be expected to be small which is determined by [116] $C_9^{eff}(\hat{s}_0) = -\frac{2m_b}{M_B \hat{s}_0} C_7^{eff}$. However, $O(\alpha_s)$ corrections to the HQET-symmetry relations lead to substantial change in the profile of the FB-asymmetry function as well as a significant shift in $\hat{s}_0$ [117,118]. They have been worked out for $B \to K^*\ell^+\ell^-$ using SCET [119]. Restricting ourselves to the kinematic region where the light K^* meson moves fast and can be viewed approximately as a collinear particle, a factorization formula for the decay amplitude of $B \to K^* \ell^+ \ell^-$, to leading power in $1/m_b$, has been derived in SCET [119]. This coincides *formally* with the formula obtained earlier by Beneke et al. [117], using the QCD factorization approach [120,121], but is valid to all orders of α_s:

$$\langle K_a^* \ell^+ \ell^- | H_{eff} | B \rangle = T_a^I(q^2)\zeta_a(q^2) + + \sum_{\pm} \int_0^\infty \frac{d\omega}{\omega} \phi_{\pm}^B(\omega) \int_0^1 du\ \phi_{K^*}^a(u) T_{a,\pm}^{II}(\omega, u, q^2) \,, \tag{34}$$

where $a = \parallel, \perp$ denotes the polarization of the K^* meson. The functions T_a^I and $T_{a,\pm}^{II}$ are perturbatively calculable. $\zeta_a(q^2)$ are the soft form factors defined in SCET while $\phi_{\pm}^B(\omega)$ and $\phi_{K^*}^a(u)$ are the light-cone distribution amplitudes (LCDAs) for the B and K^* mesons, respectively. In particular, the location of the zero of the forward-backward asymmetry in $B \to K^* \ell^+ \ell^-$, q_0^2, can be predicted more precisely in SCET due to the improved theoretical precision on the scale dependence of q_0^2.

Including the order α_s corrections, the analysis in [119] estimates the zero-point of the FB asymmetry to be

$$q_0^2 = (4.07^{+0.16}_{-0.13})\ \text{GeV}^2 \,, \tag{35}$$

of which the scale-related uncertainty is $\Delta(q_0^2)_{\text{scale}} =^{+0.08}_{-0.05}$ GeV^2 for the range $m_b/2 \le \mu_h \le 2m_b$ together with the jet function scale $\mu_l = \sqrt{\mu_h \times 0.5\ \text{GeV}}$. This is to be compared with the result given in [118], also obtained in the absence of $1/m_b$ corrections: $q_0^2 = (4.39^{+0.38}_{-0.35})$ GeV^2. Of this the largest single uncertainty (about ± 0.25 GeV^2) is attributed to the scale dependence. The difference in the estimates of the scale dependence of q_0^2 in [119] and [118] is both due to the incorporation of the SCET logarithmic resummation (done in [119]) and the different (scheme-dependent) definitions of the effective form factors for the SCET currents used in these references Power corrections in $1/m_b$ are probably comparable to the $O(\alpha_s)$ corrections, as argued in [118]. So far, q_0^2 has not been measured experimentally.

BELLE has published the first measurements [5,122] of the forward-backward asymmetry (FBA) [112]. The best-fit results by BELLE for the Wilson coefficient ratios for negative value of C_7, $\frac{C_9}{C_7} = -15.3^{+3.4}_{-4.8} \pm 1.1$ and $\frac{C_{10}}{C_7} = 10.3^{+5.2}_{-3.5} \pm 1.8$, are consistent with the SM values $C_9/C_7 \simeq -13.7$ and $C_{10}/C_7 \simeq +14.9$, evaluated in the NLO approximation. However, for the positive value of C_7, the measurements lead to $\frac{C_9}{C_7} = -16.3^{+3.7}_{-3.7} \pm 1.1$ and $\frac{C_{10}}{C_7} = +11.1^{+6.0}_{-3.9} \pm 1.8$ and the two solutions are of comparable significance. With more data at the current B factories, and yet more anticipated at the

LHC, these measurements are expected to become very precise, providing a precision test of the SM in the flavor sector.

4.3. $B \to V\gamma$: SM vs. Experiments

The decays $B \to V\gamma$ ($V = K^*, \rho, \omega$) have been calculated in the NLO approximation using the effective Lagrangian given in (23) and its analogue for $b \to d$ transitions. Two dynamical approaches, namely the QCD Factorization [120] and pQCD [123] have been employed to establish factorization of the radiative decay amplitudes in the heavy-quark limit. We illustrate the QCD-F method, where this factorization is worked out for the $B \to V\gamma$ [124,125,118,126–129] (see [130,131] for phenomenological updates in NLO, and [132,133] for the alternative "perturbative QCD" approach). In particular, the matrix element of a given operator in the effective weak Hamiltonian can be written in the form

$$\langle V\gamma\,|Q_i|\,\bar{B}\rangle = F^{B\to V_\perp}\,T_i^{\rm I} + \int d\omega\, du\, \phi_+^B(\omega)\,\phi_\perp^V(u)\,T_i^{\rm II}(\omega,u)\,. \tag{36}$$

The non-perturbative effects are contained in $F^{B\to V_\perp}$, the $B \to V$ transition form factor at $q^2 = 0$, and in ϕ_+^B and $\phi_\perp^V$, the leading-twist LCDAs of the B- and V-mesons. The hard-scattering kernels $T_i^{\rm I}$ and $T_i^{\rm II}$ include only short-distance effects and are calculable in perturbation theory. Contributions to the kernel $T^{\rm I}$ are closely related to the virtual corrections to the inclusive decay rate, and are referred to as vertex corrections. Those to the kernel $T^{\rm II}$ are related to parton exchange with the light quark in the B-meson, a mechanism commonly referred to as hard spectator scattering. It is expected that the factorization formula is valid up to corrections of $\mathcal{O}(\Lambda_{\rm QCD}/m_b)$.

The derivation of the factorization formula from a two-step matching procedure in SCET has provided additional insight into its structure. The technical details for $B \to V\gamma$ in NLO have been provided in [134,135]. In the SCET approach the factorization formula is written as

$$\langle V\gamma\,|Q_i|\,\bar{B}\rangle = \Delta_i C^A \zeta_{V_\perp} + \frac{\sqrt{m_B} F f_{V_\perp}}{4} \int d\omega\, du\, \phi_+^B(\omega)\,\phi_\perp^V(u)\,t_i^{\rm II}(\omega,u)\,, \tag{37}$$

where F and $f_{V_\perp}$ are meson decay constants. The SCET form factor $\zeta_{V_\perp}$ is related to the QCD form factor through perturbative and power corrections [117,136–138,?,140,141]. In SCET the perturbative hard-scattering kernels are the matching coefficients $\Delta_i C^A$ and $t_i^{\rm II}$. They are known completely to next-to-leading order (NLO) ($\mathcal{O}(\alpha_s)$) in renormalization-group (RG) improved perturbation theory [135]. Recently, important steps towards a complete analysis at the next-to-next-to-leading order (NNLO) in $B \to V\gamma$ decays have been derived in [142] by obtaining full results for the hard-scattering kernels for the dipole operators Q_7 and Q_8, and partial results for Q_1, valid in the large-β_0 limit and neglecting NNLO corrections from spectator scattering. In addition, this work provides the virtual corrections to this order for the $B \to V\gamma$ decays, as they can not be obtained from the published calculations for the inclusive decay $B \to X_s\gamma$, discussed previously.

In SCET the hard-scattering kernel $t_i^{\rm II}$ for a given operator is subfactorized into the convolution of a hard-coefficient function with a universal jet function, in the form

$$t_i^{\rm II}(u,\omega) = \int_0^1 d\tau \Delta_i C^{B1}(\tau) j_\perp(\tau,u,\omega) \equiv \Delta_i C^{B1} \star j_\perp . \qquad (38)$$

The hard coefficients $\Delta_i C^{B1}$ contain physics at the hard scale m_b, while the jet function $j_\perp$ contains physics at the hard-collinear scale $\sqrt{m_b\Lambda}$. The hard coefficient is identified in a first step of matching QCD $\to$ SCET$_{\rm I}$, and the jet function in a second step of matching SCET$_{\rm I}$ $\to$ SCET$_{\rm II}$. Details in NLO have been worked out for $B \to V\gamma$ in [134,135].

The effective field-theory techniques are crucial for providing a field-theoretical definition of the objects in (37), and for resumming large perturbative logarithms of the ratio $m_b/\Lambda_{\rm QCD}$ in $t_i^{\rm II}$. In the effective-theory approach resummation is carried out by solving the renormalization-group equations for the matching coefficients $\Delta_i C^{B1}$. Since these coefficients enter the factorization formula in a convolution with the jet function $j_\perp$, their anomalous dimension is a distribution in the variables τ and u. The evolution equations must be solved before performing the convolution with $j_\perp$. Therefore, resummation is not possible in the original QCD factorization formula (36), where the hard-scattering kernels $T_i^{\rm II}$ are obtained only after this convolution has been carried out.

Using the SCET framework, the branching ratios in the NNLO are as follows [142]:

$$\begin{aligned}
\mathcal{B}_{\rm NNLO}(B^0 \to K^{*0}\gamma) &\simeq (4.3 \pm 1.4) \times 10^{-5} \, , \\
\mathcal{B}_{\rm NNLO}(B^\pm \to K^{*\pm}\gamma) &\simeq (4.6 \pm 1.4) \times 10^{-5} \, , \\
\mathcal{B}_{\rm NNLO}(B_s^0 \to \phi\gamma) &\simeq (4.3 \pm 1.4) \times 10^{-5} \, .
\end{aligned}$$

It should be noted that, very much like the $B \to X_s\gamma$ case, the complete NNLO calculations for the virtual corrections to the matrix element of the operator O_1 in $B \to K^*\gamma$ are not yet at hand. In addition, the hard spectator corrections from this operator are calculated only in NLO. In the

NNLO branching ratios quoted above, the errors are increased to take these missing pieces into account.

The above theoretical branching ratios, compared with the current experimental measurements [6,7], yield the following results [142]:

$$\frac{\mathcal{B}_{\rm NNLO}(B^0 \to K^{*0}\gamma)}{\mathcal{B}_{\rm Expt}(B^0 \to K^{*0}\gamma)} = 1.1 \pm 0.35 \pm 0.06,$$
$$\frac{\mathcal{B}_{\rm NNLO}(B^\pm \to K^{*\pm}\gamma)}{\mathcal{B}_{\rm Expt}(B^\pm \to K^{*\pm}\gamma)} = 1.1 \pm 0.35 \pm 0.07,$$
$$\frac{\mathcal{B}_{\rm NNLO}(B_s^0 \to \phi\gamma)}{\mathcal{B}_{\rm Expt}(B_s^0 \to \phi\gamma)} = 0.8 \pm 0.2 \pm 0.3 \, .$$

The decays $B \to (\rho,\omega)\gamma$ involve in addition to the (short-distance) penguin amplitude also significant long-distance contributions, in particular in the decays $B^\pm \to \rho^\pm\gamma$. In the factorization approximation, typical Annihilation-to-Penguin amplitude ratio is estimated as [143]: $\epsilon_{\rm A}(\rho^\pm\gamma) = 0.30 \pm 0.07$. $O(\alpha_s)$ corrections to the annihilation amplitude in $B \to \rho\gamma$ are not known; also the proof of factorization of this amplitude is still not at hand. The annihilation contribution to the decays $B^0 \to \rho^0\gamma$ and $B^0 \to \omega\gamma$ is expected to be suppressed (relative to the corresponding amplitude in $B^\pm \to \rho^\pm\gamma$) due to the electric charges ($Q_d/Q_u = -1/2$) and the color factors, and the corresponding A/P ratio for these decays is estimated as $\epsilon_{\rm A}(\rho^0\gamma) \simeq -\epsilon_{\rm A}(\omega\gamma) \simeq 0.05$.

The decay amplitudes for $B \to (\rho,\omega)\gamma$ depend on the CKM matrix elements $V_{td}^* V_{tb}$ (from the penguin diagrams) and $V_{ub}^* V_{ud}$ (from the annihilation diagrams). Hence, these decays provide potentially very powerful constraints on the CKM parameters, $\bar{\rho}$ and $\bar{\eta}$. Since a large number of observables can be measured in these decays, such as the individual branching ratios for $B^\pm \to \rho^\pm\gamma$ and $B^0 \to (\rho^0,\omega)\gamma$, isospin- and SU(3)-violating asymmetries in the decay rates, and direct and time-dependent CP asymmetries, they have been studied theoretically in a number of papers [124,125,127,128,130,131,144]. Experimentally, a beginning has been made in the measurements of the $b \to d\gamma$ transition through the measurements of the branching ratios for $B \to (\rho,\omega)\gamma$, reported by BABAR and BELLE. Current measurements are not very precise, as can be seen from the current world averages [7] (in units of 10^{-6}): $\mathcal{B}(B^\pm \to \rho^\pm\gamma) = 0.96 \pm 0.23$, $\mathcal{B}(B^0 \to \rho^0\gamma) = 0.77 \pm 0.14$ and $\mathcal{B}(B^0 \to \omega\gamma) = 0.41 \pm 0.15$. In addition, theoretical estimates suffer from large hadronic uncertainties, dominated by the imprecise knowledge of the form factors. Hence, the resulting constraints on the CKM parameters are not very quantitative. Theoretical uncertainties are greatly reduced in the ratios of the branching ratios involving the decays $B \to (\rho,\omega)\gamma$ and $B \to K^*\gamma$. Calling the ratios of the branching ratios $R^{\pm}(\rho\gamma/K^*\gamma)$ and $R^0(\rho\gamma/K^*\gamma)$, for the decays of the $B^{\pm}$ and B^0 mesons, respectively, one has [124]

$$\begin{aligned} R^{\pm}(\rho\gamma/K^*\gamma) &= \left|\frac{V_{td}}{V_{ts}}\right|^2 \frac{(M_B^2 - M_\rho^2)^3}{(M_B^2 - M_{K^*}^2)^3}\, \zeta^2(1 + \Delta R^{\pm}(\epsilon_{\rm A}^{\pm}, \bar{\rho}, \bar{\eta}))\ , \\ R^0(\rho\gamma/K^*\gamma) &= \frac{1}{2}\left|\frac{V_{td}}{V_{ts}}\right|^2 \frac{(M_B^2 - M_\rho^2)^3}{(M_B^2 - M_{K^*}^2)^3}\, \zeta^2(1 + \Delta R^0(\epsilon_{\rm A}^0, \bar{\rho}, \bar{\eta}))\ , \end{aligned} \tag{39}$$

where $\zeta = T_1^\rho(0)/T_1^{K^*}(0)$, with $T_1^\rho(0)$ and $T_1^{K^*}(0)$ being the transition form factors evaluated at $q^2 = 0$ in the decays $B \to \rho\gamma$ and $B \to K^*\gamma$, respectively. The functions $\Delta R^{\pm}(\epsilon_{\rm A}^{\pm}, \bar{\rho}, \bar{\eta})$ and $\Delta R^0(\epsilon_{\rm A}^0, \bar{\rho}, \bar{\eta})$, appearing on the r.h.s. of the above equations encode both the $O(\alpha_s)$ contribution to the penguin amplitudes and annihilation contributions, with the latter estimated so far only in the lowest order. They have a non-trivial dependence on the CKM parameters $\bar{\rho}$ and $\bar{\eta}$ [124,125]. Theoretical uncertainty in the evaluation of the ratios $R^{\pm}(\rho\gamma/K^*\gamma)$ and $R^0(\rho\gamma/K^*\gamma)$ is dominated by the error on the quantity ζ, and to some extent also by the errors on the parameters $\epsilon_A^{\pm}$ and ϵ_A^0, characterizing the annihilation/penguin ratios. In the SU(3) limit $\zeta = 1$; SU(3)-breaking corrections have been calculated in several approaches, including the QCD sum rules and Lattice QCD. With the current values for the ratios $R^{\pm}(\rho\gamma/K^*\gamma) = 0.032 \pm 0.008$, $R^0(\rho\gamma/K^*\gamma) = 0.039 \pm 0.007$ and $R^0(\omega\gamma/K^*\gamma) = 0.021 \pm 0.007$, the current world average of $|V_{td}/V_{ts}|$ from the ratio of $B \to (\rho,\omega)\gamma$ and $B \to K^*\gamma$ is [145]:

$$|V_{td}/V_{ts}| = 0.194^{+0.015}_{-0.014}(\text{exp}) \pm 0.014(\text{th})\ , \tag{40}$$

where the Light-cone QCD sum rules [131] have been used to estimate the hadronic input quantities. This determination is compatible with the one from the mass difference ratio $\Delta M_{B_s}/\Delta M_{B_d}$ given in eq. (22), but less precise.

5. $B \to M_1 M_2$ Decays

Exclusive non-leptonic decays are the hardest nuts to crack in the theory of B-decays. Basically, there are four different theoretical approaches to calculate and/or parameterize the hadronic matrix elements in $B \to M_1 M_2$ decays:

(1) SU(2)/SU(3) symmetries and phenomenological Ansaetze [146–149]

(2) Dynamical approaches based on perturbative QCD, such as the QCD Factorization [120] and the competing pQCD approach [123].
(3) Charming Penguins [150] using the renormalization group invariant topological approach of Buras and Silvestrini [151].
(4) Soft Collinear Effective Theory (SCET) [43–46], which we have already discussed in the context of radiative and semileptonic decays.

These approaches will be discussed on the example of the $B \to \pi\pi$ and $B \to K\pi$ decays for which now there exist enough data to extract the underlying dynamical parameters. Prior to this, however, we discuss the measurements of the angle β (or ϕ_1) from the experiments at the B-factories.

5.1. *Interplay of Mixing and Decays of B^0- and $\bar{B}^0$-Mesons to CP Eigenstates*

We start with the discussion of the transition $b \to c\bar{c}\bar{s}$, which is dominated by the tree topology. The time-dependent CP asymmetry in the decays $B^0 \to f$ and $\bar{B}^0 \to f$, where f is a CP eigenstate, such as $J/\psi K_s$ and $J/\psi K_L$, is defined as:

$$\mathcal{A}_f(t) = \frac{\Gamma[\bar{B}^0(t) \to f] - \Gamma[B^0(t) \to f]}{\Gamma[\bar{B}^0(t) \to f] + \Gamma[B^0(t) \to f]}. \tag{41}$$

The time evolution of the two flavor eigenstates B^0 and $\bar{B}^0$ is determined by (2×2) Hermitian matrices M and Γ. The physical states (with definite masses and lifetimes) are the linear combinations of B^0 and $\bar{B}^0$, with $|B_d(L, H)\rangle = p|B^0\rangle \pm q|\bar{B}^0\rangle$, dependent on two complex parameters p and q. Defining the decay amplitudes $A(f) \equiv \langle f|H|B^0\rangle$ and $\bar{A}(f) \equiv \langle f|H|\bar{B}^0\rangle$ of the B^0- and $\bar{B}^0$-mesons into the final state f, the time-dependent CP asymmetry is determined by the quantity λ_f, involving the interplay of mixing and decay amplitudes:

$$\lambda_f = \frac{q}{p}\,\rho(f), \qquad \rho(f) = \frac{\bar{A}(f)}{A(f)}. \tag{42}$$

For the B_d^0 - $\bar{B}_d^0$ mixing, the ratio q/p involves the phase β (or ϕ_1), which is one of the angles of the unitarity triangle:

$$\frac{q}{p} = \frac{V_{tb}^* V_{td}}{V_{tb} V_{td}^*} = \mathrm{e}^{-2i\phi_{\rm mixing}} = \mathrm{e}^{-2i\beta}. \tag{43}$$

The time dependent CP asymmetry (41) is then expressed as

$$\mathcal{A}_f(t) = C_f \cos(\Delta M_{B_d} t) + S_f \sin(\Delta M_{B_d} t), \tag{44}$$

where $\Delta M_{B_d} = (0.507 \pm 0.005)$ ps^{-1} is the mass difference between the heavy and light B_d^0-meson mass eigenstates and the difference in the decay widths $\Delta\Gamma_{B_d}$ has been neglected. The quantities C_f and S_f, called the direct and mixing-induced CP asymmetries, respectively, are defined in terms of the complex variable λ_f as follows:

$$C_f = \frac{1 - |\lambda_f|^2}{1 + |\lambda_f|^2}, \qquad S_f = \frac{2\text{Im}\lambda_f}{1 + |\lambda_f|^2}. \tag{45}$$

If the decays $B^0 \to f$ and $\bar{B}^0 \to f$ are dominated by a *single* amplitude, the ratio $\rho(f) = \eta_f \, e^{-2i\phi_{\text{decay}}}$, where $\eta_f = \pm 1$ is the CP parity of the state f, is a pure phase factor and the asymmetries (45) reduce to the expressions:

$$C_f = 0, \qquad S_f = -\eta_f \sin 2(\phi_{\text{mixing}} + \phi_{\text{decay}}). \tag{46}$$

The decays $B^0/\bar{B}^0 \to J/\psi K_s, J/\psi K_L$, and a number of related final states with f being $\psi(2S)K_s$, $\eta_c K_s$, $\chi_{c1}K_s$, and $J/\psi K^{*0}(K^{*0} \to K_s\pi^0)$ belong to the category of *gold plated* decays [152]. In all these modes, the direct CP asymmetry C_f, to a very high accuracy, vanishes, and the quantity S_f, the mixing-induced CP asymmetry, measures $\sin(2\beta)$. Averaging over all the decay channels, the results of the BABAR and BELLE measurements are as follows [4]:

$$C = 0.049 \pm 0.022 \pm 0.017, \quad S = 0.714 \pm 0.032 \pm 0.018, \quad \text{[BABAR]} \tag{47}$$

$$C = -0.019 \pm 0.025, \quad S = 0.651 \pm 0.034. \quad \text{[BELLE]} \tag{48}$$

In the BABAR result, the first error is statistical and the second is systematic, while in the BELLE data both the errors have been combined. The current world average for $S_f = \sin(2\beta)$ for the quark transition $b \to c\bar{c}s$ is [7]:

$$\sin(2\beta) = 0.681 \pm 0.025, \tag{49}$$

where the data from LEP and Tevatron have also been included. Restricting β in the range $0 \leq \beta \leq \pi/2$, two possible values can be extracted $\beta = (21.5 \pm 1.0)^\circ$ and $\beta = (68.5 \pm 1.0)^\circ$. The two-fold ambiguity has now been resolved by several $\cos(2\beta)$ measurements, involving the Dalitz analysis of the decay modes $B^0 \to D^0_{3-\text{body}}h^0$, $B^0 \to K_s\pi^+\pi^-$, $B^0 \to K_s K^+K^-$, and the older results on $B^0 \to J/\psi K^{*0}$, leading to the determination $\beta = (21.5 \pm 1.0)^\circ$.

The *direct* measurement of $\sin(2\beta)$ in eq.(49) is to be compared with the *indirect* estimate of the same, obtained from the fits of the CKM unitarity triangle (UT). For this, the UTfit collaboration [66] quotes $\sin(2\beta) = 0.739 \pm 0.044$, obtained from the sides of the UT alone, and $\sin(2\beta) = 0.736 \pm$

0.042, by including also the CP-violating quantity ϵ_K in the K-decays. The results from the CKMfitter group [67] are similar. Thus, SM passes this test comfortably.

Another key test of the SM in the flavor sector is to compare the CP-violating quantities S_f and C_f involving the penguin-topology dominated quark transitions $b \to s\bar{s}s$ and $b \to sd\bar{d}$ with the ones from the transition $b \to c\bar{c}s$, dominated by the tree topology and discussed quantitatively above. The point here is that penguin amplitudes may receive contributions from New Physics. For example, new phases, present generically in supersymmetric theories, may reveal themselves, leading to $S_{f=c\bar{c}s} \neq S_{f=s\bar{s}s;sd\bar{d}}$. Examples of the final states induced by the transition $b \to s\bar{s}s$ are $(\phi, \eta, \eta', K\bar{K})K_s$, and the ones induced by the transition $b \to sd\bar{d}$ are $K_s^0(\pi^0, \rho^0, \omega)$ ($B^0 \to \eta' K_s$ receives contributions from both the transitions). The current measurements of $S_f = -\eta_f \sin(2\beta^{\text{eff}})$ from the penguin-dominated decays are [7]: $S_{\phi K^0} = 0.39 \pm 0.17$, $S_{\eta' K^0} = 0.61 \pm 0.07$, $S_{K_s K_s K_s} = 0.58 \pm 0.20$, $S_{\pi^0 K_s} = 0.38 \pm 0.19$, $S_{\rho^0 K_s} = 0.61^{+0.25}_{-0.27}$, $S_{\omega K_s} = 0.48 \pm 0.24$, $S_{f_0 K^0} = 0.84 \pm 0.07$, $S_{K^+K^-K^0} = 0.73 \pm 0.10$, and $S_{\pi^0\pi^0 K_s} = -0.52 \pm 0.41$. These measurements are not as precise as the ones from the $b \to c\bar{c}s$ decays due to the much smaller branching ratios (typically 10^{-5}) compared to the decay $B^0 \to J/\psi K_s$. Also, they involve more than one decay topologies. However, within (large) errors, the values of $\sin(2\beta^{\text{eff}})$ from the penguin-dominated transitions are consistent with the value of $\sin(2\beta)$ from the tree-dominated transition given above in eq.(49), with the possible exception of $S_{\phi K^0}$, which deviates by about 2σ, and the poorly measured odd-man out $S_{\pi^0\pi^0 K_s}$. It seems that the fog on S_f in the penguin-dominated decays from the initial epoch of the B-factory experiments has largely evaporated, and the emerging contours of CP asymmetries in these decays are very much the same as predicted by the SM.

5.2. $B \to \pi\pi$: *SM vs. Experiments*

The determination of the phase α is based on the branching ratios and CP asymmetries in the quark transition $b \to u\bar{u}d$. They metamorphise in the decays $B \to \pi\pi$, $B \to \rho\pi$ and $B \to \rho\rho$, apart from other final states. We concentrate here on the decay $B \to \pi\pi$, which has received a lot of theoretical attention.

There are three dominant topologies in the $B \to \pi\pi$ decays termed as Tree (T), Penguin (P) and Color-suppressed (C). In addition, there are several other subdominant topologies which will be neglected in the discussion below. Parametrization of the T, P, and C amplitudes is convention-dependent. In the Gronau-Rosner c-convention [149], these amplitudes can be represented as

$$\begin{aligned}\sqrt{2}\,A^{+0} &= -|T|\,\mathrm{e}^{i\delta_T}\,\mathrm{e}^{i\gamma}\left[1+|C/T|\,\mathrm{e}^{i\Delta}\right] ,\\ A^{+-} &= -|T|\,\mathrm{e}^{i\delta_T}\left[\mathrm{e}^{i\gamma}+|P/T|\,\mathrm{e}^{i\delta}\right] ,\\ \sqrt{2}\,A^{00} &= -|T|\,\mathrm{e}^{i\delta_T}\left[|C/T|\,\mathrm{e}^{i\Delta}\,\mathrm{e}^{i\gamma}-|P/T|\,\mathrm{e}^{i\delta}\right] .\end{aligned} \tag{50}$$

The charged-conjugate amplitudes $\bar{A}^{ij}$ differ by the replacement $\gamma \to -\gamma$. The amplitudes (50) and the charged-conjugate ones obey the isospin relations:

$$A^{+0} = \frac{1}{\sqrt{2}}\,A^{+-} + A^{00}, \qquad \bar{A}^{-0} = \frac{1}{\sqrt{2}}\,\bar{A}^{+-} + \bar{A}^{00}. \tag{51}$$

There are 5 dynamical parameters $|T|$, $r \equiv |P/T|$, δ, $|C/T|$, Δ, with $\delta_T = 0$ assumed for the overall phase. Thus, the weak phase γ can be extracted together with other quantities if the complete set of experimental data on $B \to \pi\pi$ decays is available.

Table 1. Branching ratios (in units of 10^{-6}) and CP asymmetries in the $B \to \pi\pi$ decays

$\mathcal{B}(B^+ \to \pi^+\pi^0) = 5.59^{+0.41}_{-0.40}$	$A_{\rm CP}(\pi^+\pi^0) = 0.06 \pm 0.05$
$\mathcal{B}(B^0 \to \pi^0\pi^0) = 1.31 \pm 0.21$	$A_{\rm CP}(\pi^0\pi^0) = 0.48^{+0.32}_{-0.31}$
$\mathcal{B}(B^0 \to \pi^+\pi^-) = 5.16 \pm 0.22$	$C_{\rm CP}(\pi^+\pi^-) = -0.38 \pm 0.07$
	$S_{\rm CP}(\pi^+\pi^-) = -0.61 \pm 0.08$

The experimental branching ratios and the direct CP asymmetries $A_{\rm CP}(\pi^0\pi^0)$ and $C_{\rm CP}(\pi^+\pi^-)$, as well as the value of the coefficient $S_{\rm CP}(\pi^+\pi^-)$ in time-dependent CP asymmetry, presented in Table 1, have been fitted to determine the various parameters (the direct CP asymmetry $A_{\rm CP}(\pi^+\pi^0)$ is not relevant for this analysis but can be important in determining the size of electroweak contribution in the decays considered). An updated analysis by Parkhomenko based on the paper [153] yields the following values for the hadronic parameters:

$$\begin{aligned}|P/T| &= 0.473^{+0.060}_{-0.055}, \qquad \delta = (-40.2^{+6.8}_{-4.7})^\circ ,\\ |C/T| &= 0.966^{+0.058}_{-0.061}, \qquad \Delta = (-56.3^{+8.4}_{-7.9})^\circ,\end{aligned} \tag{52}$$

and for the CKM unitarity triangle angle γ (or equivalently α)

$$\gamma = \left(65.9^{+3.0}_{-3.2}\right)^\circ, \qquad \alpha = \pi - \beta - \gamma = \left(92.6^{+3.4}_{-3.2}\right)^\circ . \tag{53}$$

Similar fits based on their data have been performed by the BABAR and BELLE collaborations resulting in slightly larger values: $\alpha = (96^{+10}_{-6})^\circ$ (BABAR) and $\alpha = (97 \pm 11)^\circ$ (BELLE). The overall fits performed by the CKM-Fitter and UT-Fit groups prefer slightly smaller values, yielding: $\alpha = (90.7^{+4.5}_{-2.9})^\circ$ [67] and $\alpha = (88.7 \pm 6.2)^\circ$ [66], respectively. All the above estimates are in good agreement with each other within the quoted errors, stating that the data on $B \to \pi\pi$ (as well as the other decay modes $B \to \rho\pi$ and $B \to \rho\rho$) are in agreement with the indirect estimate of the phase α from the unitarity triangle. The strong phases δ and Δ in Eq.(52) come out rather large. In particular, they are much larger than the predictions of the QCD-F approach [120], with pQCD [123] in better agreement with data, but neither of these approaches provides a good fit of the entire $B \to \pi\pi$ data.

Data on $B \to \pi\pi$ decays are in agreement with the phenomenological approach of the so-called charming penguins [154], and with the SCET-based analyses by Bauer et al. [155,156] which also attributes a dominant role to the charming penguin amplitude. However, a proof of the factorization of the charming penguin amplitude in the SCET approach remains to be provided. In addition, SCET makes a number of predictions in the $B \to \pi\pi$ sector, such as the branching ratio $\mathcal{B}(B^0 \to \pi^0\pi^0)$: [155]

$$\mathcal{B}(B^0 \to \pi^0\pi^0)\Big|_{\gamma=64^\circ} = (1.3 \pm 0.6) \times 10^{-6} \,. \tag{54}$$

In contrast, predictions of the QCD-F and pQCD approaches are rather similar: $\mathcal{B}(B^0 \to \pi^0\pi^0) \sim 0.3 \times 10^{-6}$, in substantial disagreement with the data.

5.3. *Present bounds on the phase γ from B decays*

The classic method for determining the phase γ (or ϕ_3) [157–160] involves the interference of the tree amplitudes $b \to uW^- \to u\bar{c}s$ leading to $B^- \to D^0K^-$ and $b \to cW^- \to c\bar{u}s$ leading to $B^- \to \overline{D^0}K^-$. These amplitudes can interfere if D^0 and $\overline{D^0}$ decay into a common hadronic final state. Noting that the CP$= \pm 1$ eigenstates $D^0_\pm$ are linear combinations of the D^0 and $\overline{D^0}$ states: $D^0_\pm = (D^0 \pm \overline{D^0})/\sqrt{2}$, both branches lead to the same final states $B^- \to D^0_\pm K^-$. So, the condition of CP interferometry is fulfilled. The decays $B^- \to D^0_\pm K^-$ are described by the amplitudes:

$$A(B^- \to D^0_\pm K^-) = \frac{1}{\sqrt{2}}\left[A(B^- \to D^0K^-) \pm A(B^- \to \overline{D^0}K^-)\right] . \tag{55}$$

Since, the weak phase of the $b \to u$ transition is γ but the $b \to c$ transition has no phase, a measurement of the CP asymmetry through the interference of these two amplitudes yields γ. The four equations that will be used to extract γ are:

$$R_\pm \equiv \frac{\mathcal{B}(B^- \to D^0_\pm K^-) + \mathcal{B}(B^+ \to D^0_\pm K^+)}{\mathcal{B}(B^- \to D^0 K^-) + \mathcal{B}(B^+ \to D^0 K^+)} = 1 + r^2_{\rm DK} \pm 2r_{\rm DK}\cos\delta_{\rm DK}\cos\gamma\,,$$

$$A_\pm \equiv \frac{\mathcal{B}(B^- \to D^0_\pm K^-) - \mathcal{B}(B^+ \to D^0_\pm K^+)}{\mathcal{B}(B^- \to D^0_\pm K^-) + \mathcal{B}(B^+ \to D^0_\pm K^+)} = \frac{\pm 2r_{\rm DK}\sin\delta_{\rm DK}\sin\gamma}{1 + r^2_{\rm DK} \pm 2r_{\rm DK}\cos\delta_{\rm DK}\cos\gamma}\,. \tag{56}$$

Here, $r_{\rm DK}$ is the ratio of the two tree amplitudes [161] $r_{\rm DK} \equiv |T_1/T_2| \sim (0.1 - 0.2)$, with T_1 and T_2 being the CKM suppressed ($b \to u$) and CKM allowed ($b \to c$) amplitudes, respectively, and $\delta_{\rm DK}$ is the relative strong phase between them. The construction of the final states involves flavor and CP-tagging of the various D^0 states, which can be done, for example, through the decays $D^0_+ \to \pi^+\pi^-$, $D^0_- \to K_S\pi^0$, and $D^0 \to K^-\pi^+$. With three unknowns ($r_{\rm DK}, \delta_{\rm DK}, \gamma$), but four quantities which will be measured, $R_\pm$ and $A_\pm$, one has, in principle, an over constrained system.

Experimentally, the quantities $R_\pm$ are measured through the ratios:

$$R(K/\pi) \equiv \frac{\mathcal{B}(B^- \to D^0 K^-)}{\mathcal{B}(B^- \to D^0 \pi^-)}\,, \qquad R(K/\pi)_\pm \equiv \frac{\mathcal{B}(B^\pm \to D^0_\pm K^\pm)}{\mathcal{B}(B^\pm \to D^0_\pm \pi^\pm)}\,. \tag{57}$$

With all three quantities $R(K/\pi)$ and $R(K/\pi)_\pm$ measured, one can determine $R_\pm = R(K/\pi)_\pm / R(K/\pi)$. More useful decay modes to construct the $B \to DK$ triangle can be added to reduce the statistical errors. Along these lines, Atwood and Soni [162] have advocated to also include the decays of the vector states in the analysis, such as $B^- \to K^{*-}D^0$, $B^- \to K^- D^{*0}$, and $B^- \to K^{*-}D^{*0}$, making use of the $D^{*0} \to D^0\gamma$ and $D^{*0} \to D^0\pi^0$ modes.

Present measurements in the $B \to DK$ and $B \to D\pi$ decays by the BABAR and BELLE collaborations yielding $R_\pm$ and $A_\pm$ for the $D_{\rm CP}K^-$ mode are summarized by HFAG [7]:

$$R_+ = 1.09 \pm 0.09\,, \qquad A_+ = 0.26 \pm 0.08 \;\; [\text{BELLE, BABAR}]\,,$$
$$R_- = 0.90 \pm 0.10\,, \qquad A_- = -0.16 \pm 0.09 \;\; [\text{BELLE}]\,.$$

The corresponding quantities for the $D^*_{\rm CP}K^-$ and $D_{\rm CP}K^{*-}$ are also given by HFAG [7].

A modification of the Gronau-London-Wyler (GLW) method described above has been suggested by Atwood, Dunietz and Soni (ADS), where $B^- \to D^0 K^-$ with $D^0 \to K^+\pi^-$ (or similar) and the charge conjugate decays are implied. BABAR and BELLE use the following definitions for the quantities called $A_{\rm ADS}$ and $R_{\rm ADS}$, (the decay modes $B^- \to D^0K^-$ followed by $D^0 \to K^+\pi^-$ are used to exemplify the method)

$$\begin{aligned} R_{\rm ADS} &\equiv \frac{\mathcal{B}(B^- \to [K^+\pi^-]_D K^-) + \mathcal{B}(B^+ \to [K^-\pi^+]_D K^+)}{\mathcal{B}(B^- \to [K^-\pi^+]_D K^-) + \mathcal{B}(B^+ \to [K^+\pi^-]_D K^+)}\,, \\ A_{\rm ADS} &\equiv \frac{\mathcal{B}(B^- \to [K^+\pi^-]_D K^-) - \mathcal{B}(B^+ \to [K^-\pi^+]_D K^+)}{\mathcal{B}(B^- \to [K^+\pi^-]_D K^-) + \mathcal{B}(B^+ \to [K^-\pi^+]_D K^+)}\,. \end{aligned} \quad (58)$$

The current measurements of these observables are summarized by HFAG [7]. In the analysis of data, usually the GLW and ADS methods are combined and a χ^2-fit is done to determine the profile of the phase γ.

A variant of the $B \to DK$ method of measuring γ is to use the decays $B^\pm \to DK^\pm$ followed by multi-body decays of the D-meson, such as $D^0 \to K_S\pi^-\pi^+$, $D^0 \to K_SK^-K^+$ and $D^0 \to K_S\pi^-\pi^+\pi^0$, in which a binned Dalitz plot analysis of the decays $D^0/\overline{D^0} \to K_S\pi^-\pi^+$ was proposed [160,163]. Assuming no CP asymmetry in D^0 decays, the amplitude of the $B^+ \to D^0K^+ \to (K_S\pi^+\pi^-)K^+$ can be written as

$$M_+ = f(m_+^2, m_-^2) + r_{DK} e^{i(\gamma+\delta_{DK})} f(m_-^2, m_+^2)\,, \quad (59)$$

where m_+^2 and m_-^2 are the squared invariant masses of the $K_S\pi^+$ and $K_S\pi^-$ combinations in the D^0 decay, and f is the complex amplitude of the decay $D^0 \to K_S\pi^+\pi^-$. The quantities r_{DK} and δ_{DK} are the relative magnitudes and strong phases of the two amplitudes, already discussed earlier. The amplitude for the charge conjugate B^- decay is

$$M_- = f(m_-^2, m_+^2) + r_{DK} e^{i(-\gamma+\delta_{DK})} f(m_+^2, m_-^2)\,. \quad (60)$$

Once the functional form of f is fixed by a choice of a model for $D^0 \to K_S\pi^+\pi^-$ decay, the Dalitz distribution for B^+ and B^- decays can be fitted simultaneously by the expressions for M_+ and M_-, with r_{DK}, δ_{DK} and γ (or ϕ_3) as free parameters. The model-dependence could be removed by a binned Dalitz distribution [163]. This is usually called the GGSZ method, and has been used to determine γ. The combined fit of both of these methods by CKMfitter [67] yields $\gamma = (76.8^{+30.4}_{-31.5})^\circ$, to be compared with their overall fit from the CKM unitarity $\gamma = (67.6^{+2.8}_{-4.5})^\circ$. The corresponding fit by the UTfit group [66] yields $\gamma = (67 \pm 7)^\circ$. Thus, we see that within the current experimental error of the *direct* measurements, which is quite large, also the phase γ is compatible with its *indirect* estimates in the SM. The experimental precision will greatly improve at the LHC, in particular, by using 2-body B_s-decays.

5.4. $B \to K\pi$: SM vs. Experiments

We now discuss the decays $B \to K\pi$. First, we note that the direct CP-asymmetry in the $B \to K\pi$ decays has now been measured by the BABAR, BELLE and CDF collaborations:

$$A_{\rm CP}(\pi^+K^-) = \begin{cases} (-10.7 \pm 1.8^{+0.7}_{-0.4})\% & [\text{BABAR}], \\ (-9.3 \pm 1.8 \pm 0.8)\% & [\text{BELLE}], \\ (-8.6 \pm 2.3 \pm 0.9)\% & [\text{CDF}], \end{cases} \tag{61}$$

to be compared with the predictions of the two factorization-based approaches: $A_{\rm CP}(\pi^+K^-) = (-12.9 \div -21.9)\%[\text{pQCD}]$ [123] and $A_{\rm CP}(\pi^+K^-) = (-5.4 \div +13.6)\%[\text{QCD}-\text{F}]$ [120], with the latter falling short of a satisfactory description of data.

The charged and neutral $B \to \pi K$ decays have received a lot of theoretical attention. In particular, many ratios involving these decays have been proposed to test the SM [165–168] and extract useful bounds on the angle γ, starting from the Fleischer-Mannel bound [165]:

$$\sin^2\gamma \le R \equiv \frac{\tau_{B^+}}{\tau_{B^0_d}} \frac{\mathcal{B}(B^0_d \to \pi^-K^+) + \mathcal{B}(\bar{B}^0_d \to \pi^+K^-)}{\mathcal{B}(B^+ \to \pi^+K^0) + \mathcal{B}(B^- \to \pi^-\bar{K}^0)} \,. \tag{62}$$

The current experimental average $R = 0.899 \pm 0.049$ allows to put a bound: $\gamma < 92^\circ$ (at 95% C.L.). This is in agreement with the determination of γ from the $B \to \pi\pi$ and $B \to D^{(*)}K^{(*)}$ decays given earlier and the indirect unitarity constraints. Thus, both R and $A_{\rm CP}(\pi^+K^-)$ are in agreement with the SM. The same is the situation with the Lipkin sum rule [167]:

$$R_L \equiv 2\,\frac{\Gamma(B^+ \to K^+\pi^0) + \Gamma(B^0 \to K^0\pi^0)}{\Gamma(B^+ \to K^0\pi^+) + \Gamma(B^0 \to K^+\pi^-)} = 1 + \mathcal{O}(\frac{P_{\rm EW} + T}{P})^2 \,, \tag{63}$$

implying significant electroweak penguin contribution in case R_L deviates significantly from unit. With the current experimental average $R_L = 1.071 \pm 0.049$, this is obviously not the case. This leaves then the two other ratios R_c and R_n involving the $B \to \pi K$ decays of $B^\pm$ and B^0 mesons:

$$R_c \equiv 2\,\frac{\mathcal{B}(B^\pm \to \pi^0K^\pm)}{\mathcal{B}(B^\pm \to \pi^\pm K^0)} \,, \qquad R_n \equiv \frac{1}{2}\,\frac{\mathcal{B}(B^0_d \to \pi^\mp K^\pm)}{\mathcal{B}(B^0_d \to \pi^0K^0)} \,. \tag{64}$$

Their experimental values $R_c = 1.117 \pm 0.071$ and $R_n = 0.980 \pm 0.067$ are to be compared with the current SM-based estimates [164] $R_c = 1.14 \pm 0.05$ and $R_n = 1.11^{+0.04}_{-0.05}$. This implies $R_c(\text{SM}) - R_c(\text{Exp}) = 0.02 \pm 0.09$ and $R_n(\text{SM}) - R_n(\text{Exp}) = 0.13 \pm 0.08$. Possible deviations from the SM, if confirmed, would imply new physics, advocated in this context, in particular, by Yoshikawa [169], Beneke and Neubert [170] and Buras et al. [164]. However, as of now, one has to conclude that SM is in agreement with the measurements of both R_c and R_n.

Finally, a bound on $\mathcal{B}(B^0 \to K^0\overline{K^0})$ based on $SU(3)$ and $B \to \pi\pi$ data, obtained recently by Fleischer and Recksiegel [171], yielding $\mathcal{B}(B^0 \to K^0\bar{K}^0) < 1.5 \times 10^{-6}$ is well satisfied by the current world average [7] $\mathcal{B}(B^0 \to K^0\bar{K}^0) = (0.96^{+0.21}_{-0.19}) \times 10^{-6}$.

6. B_s^0 Physics: Eldorado for the Tevatron and the LHC

The main goal of b physics at the hadron colliders Tevatron and the LHC is to chart out the physics of the B_s^0 and $B_c^\pm$ mesons and of the b-baryons. The current information on the spectroscopic and decay characteristics of these hadrons is still very much in the offing, though clearly the two Tevatron experiments have made some incisive inroads in these otherwise uncharted territories. Despite the overwhelming performance of the B factory experiments, there still remain a few landmark measurements to be carried out involving B_d^0 and $B^\pm$ mesons. These include, precise measurements of the CP asymmetries in the penguin-dominated exclusive decays, quantitative determinations of the Wilson coefficients in the effective theory for weak decays (C_7, C_8, C_9, C_{10}), which will be made possible by the precise measurements of the radiative and semileptonic decays $B \to (X_s, K^*)\gamma$ and $B \to (X_s, K, K^*)\ell^+\ell^-$. It is challenging to measure the inclusive decays at the LHC, but certainly exclusive decays will be well measured.

In this section, a brief list of some selected b physics topics to be studied at the LHC is given and discussed.

- B_s^0 - $\overline{B_s^0}$ Mixing.

Apart from the precise measurement of $\Delta M_{B_s} = (17.77 \pm 0.10 \pm 0.07)$ $(\text{ps})^{-1}$ by the CDF collaboration, there are two other quantities still to be measured in this complex: Lifetime difference $\Delta\Gamma_{B_s}$ and the phase ϕ_s. These quantities have been calculated to a high precision in the SM [172]. A recent update of this work yields [71]

$$\frac{\Delta\Gamma_{B_s}}{\Delta M_{B_s}} = (49.7 \pm 9.4) \times 10^{-4}; \quad \Delta\Gamma_{B_s} = (0.096 \pm 0.039)(\text{ps})^{-1},$$
$$\phi_s = (4.2 \pm 1.4) \times 10^{-3} = 0.24^\circ \pm 0.08^\circ \ . \tag{65}$$

The current measurements of these quantities from the D0 collaboration are [173]

$$\Delta\Gamma_{B_s} = (0.12^{+0.08}_{-0.10} \pm 0.02)(\text{ps})^{-1} \ (\text{assuming } \phi_s = 0) \ , \tag{66}$$

and

$$\Delta\Gamma_{B_s} = (0.17 \pm 0.09 \pm 0.02)(\text{ps})^{-1} \; ; \phi_s = -0.79 \pm 0.56^{+0.14}_{-0.01} \; . \qquad (67)$$

The corresponding measurement (assuming $\phi_s = 0$) for $\Delta\Gamma_{B_s}$ from CDF is [174] $\Delta\Gamma_{B_s} = (0.076^{+0.059}_{-0.063} \pm 0.006)\ (\text{ps})^{-1}$, where the first error is statistical and the second systematic. At the LHCb [12], one anticipates a statistical sensitivity of $\sigma(\sin\phi_s) \sim 0.031$ and $\sigma(\Delta\Gamma_s/\Gamma_s) \sim 0.011$, assuming an integrated luminosity of 2 $(\text{fb})^{-1}$ and using the decay $B_s \to J/\psi\phi$. This sensitivity will be improved by accumulating more data and adding the CP modes $B_s \to J/\psi\phi$ and the pure CP modes $B_s \to J/\psi\eta$ and $B_s \to J/\psi\eta'$. The ATLAS and CMS sensitivities on ϕ_s are expected to be somewhat worse by typically a factor 2. Thus, experiments at the LHC will be able to test the SM estimates for both the quantities $\Delta\Gamma_{B_s}$ and ϕ_s.

- Precise measurement of the phase γ.

At the LHC one can combine the methods of determining the phase γ already in use at the B factories involving the decays $B^+ \to D^{(*)}K^+$ and $B^0 \to DK^{(*)}$, with the decay $B_s^0 \to D_s K$. In addition, one can use the U-spin symmetry arguments advocated, in particular, by Fleischer [175], to combine data from the decays $B_d^0 \to \pi^+\pi^-$ and $B_s^0 \to K^+K^-$ to constrain γ. An educated guess [176] is that a sensitivity $\sigma(\gamma) \simeq 4°$ with 2 $(\text{fb})^{-1}$ of data is reachable at the LHCb, improving to $\sigma(\gamma) \simeq 2.4°$ with 10 $(\text{fb})^{-1}$. This will result in an order of magnitude improvement over the current precision on this phase. Modest improvements are also anticipated for the other two phases α and β at the LHC.

- Leptonic decay $B_s^0 \to \mu^+\mu^-$

New and improved upper limits have been presented by the CDF [177] and D0 [178] collaborations for the decays $B_s^0 \to \mu^+\mu^-$ and $B_d^0 \to \mu^+\mu^-$. They are as follows (at 95% C.L.)

$$\begin{aligned} \mathcal{B}(B_s^0 \to \mu^+\mu^-) &< 9.3\ [5.8] \times 10^{-8} \ \ \text{D0[CDF]} \ , \\ \mathcal{B}(B_d^0 \to \mu^+\mu^-) &< 1.8 \times 10^{-8} \ \ [\text{CDF}]. \end{aligned} \qquad (68)$$

The CDF and DO upper limits have been combined to yield [5] $\mathcal{B}(B_s^0 \to \mu^+\mu^-) < 4.5 \times 10^{-8}$, to be compared with the SM predictions [179] $\mathcal{B}(B_s^0 \to \mu^+\mu^-) = 3.4 \times 10^{-9}$ and $\mathcal{B}(B_d^0 \to \mu^+\mu^-) = 1.0 \times 10^{-10}$ with $\pm 15\%$ theoretical uncertainty. Hence, currently there is no sensitivity for the SM decay rate. However, as the leptonic branching ratios probe the Higgs sector in beyond-the-SM scenarios, such as supersymmetry, and they depend sensitively on $\tan\beta$, the Tevatron upper limit on $\mathcal{B}(B_s^0 \to \mu^+\mu^-)$ probes the large $\tan\beta$ parameter space, though the precise constraints are model dependent [180,181]. At the LHC, the two main collider experiments ATLAS and CMS will reach the SM sensitivity, certainly with the higher LHC luminosity, $L_{\text{LHC}} = 10^{34}\ cm^{-2}\ \text{s}^{-1}$, as the decay $B_s^0 \to \mu^+\mu^-$ remains triggerable with the high luminosity.

- Charmless non-leptonic $B_s \to h_1 h_2$ decays.

The experimental program to study non-leptonic decays $B_s^0 \to h_1 h_2$ has started (here $h_{1,2}$ stand for charmless light vector or pseudoscalar mesons) with first measurements for the branching ratios $B_s^0 \to K^+\pi^-$ and $B_s^0 \to K^+K^-$ made available recently by the CDF collaboration [9,10]. Remarkably, the first direct CP asymmetry involving the decay $B_s^0 \to K^+\pi^-$ and its CP conjugate mode reported by CDF is found to be large, with $A_{\text{CP}}(\overline{B_s^0} \to K^+\pi^-) = (39 \pm 15 \pm 8)\%$. This large CP asymmetry was predicted by Lipkin [182] based on SU(3) symmetry arguments. This already tests various dynamical models, such as QCDF [183], SCET [184] and pQCD [185]. With the ongoing b-physics program at the Tevatron, but, in particular, with the onset of the LHC, we expect a wealth of data involving the decays of the hitherto less studied B_s^0 meson. The charmless $B_s^0 \to h_1 h_2$ decays are also important for the CP asymmetry studies and the determination of the inner angles of the unitarity triangle. As already stated, a number of charmless decays $B_s^0 \to h_1 h_2$ can be related to the $B_d^0 \to h_1 h_2$ decays using SU(3) (or U-spin) symmetry, and hence data on these decays can be combined to test the underlying standard model and search for physics beyond the SM under less (dynamical) model-dependent conditions. Anticipating the experimental developments, many studies have been devoted to the interesting charmless $B_s^0 \to h_1 h_2$ decays, waiting to be tested at the LHC.

7. Summary and Outlook

Summarizing, dedicated experiments carried out over several decades combined with progress in theoretical techniques embedded in QCD have enabled a precise determination of the CKM matrix elements. The knowledge of the third row and the third column of V_{CKM} has come from b-physics, which we discussed at length in this review. Of these, precise determination of V_{cb} and V_{ub} required good control over the perturbative and non-perturbative aspects of QCD. The current precision on the direct determination of V_{tb} from the decay $t \to bW$ is limited by statistics and this will vastly improve from the top quark studies at the LHC and later at the ILC. The determinations of V_{ts} and V_{td} require not only precise knowledge of QCD in b decays but implicitly also the assumption of the CKM unitarity, as they are determined from the loop-induced $b \to s$ and $b \to d$ transitions. Their current best measurement is through the mass differences ΔM_{B_s} and ΔM_{B_d}, and the precision on these matrix elements (typically 10%) is completely dominated by theory. A complementary determination of V_{td} and V_{ts} is also at hand from the radiative penguin transitions $b \to (d,s)\gamma$ and the exclusive decays $B \to (K^*, \rho, \omega)\gamma$, but the current precision is limited by both experimental statistics and non-perturbative aspects of QCD, such as the transition form factors in exclusive decays. This surely will improve over the next several years.

Experiments have also firmly established the phenomenon of CP violation in the K and B meson sectors. The various CP asymmetries in these decays are found compatible with each other and, with some help from QCD, have a consistent interpretation in terms of the single complex phase of the CKM matrix. Again, in principle, there is ample room also for beyond-the-SM weak phases, which would lead to very different patterns of CP asymmetries in the tree-dominated versus loop-dominated transitions. This has not been borne out by experiments at the B factories. While the current data is not equivocal on all the decay channels, and the dynamical aspects of not all the measured B-meson decays are quantitatively understood, the experimental case for the extra weak-phases is rather weak. Ongoing experiments at the B factories are expected to significantly reduce the errors on the quantities S_f and C_f in penguin-dominated decays to settle the issue of new weak phases in B decays.

From the foregoing one has to tentatively conclude that the CKM paradigm is now firmly established as the dominant mechanism of flavor transitions in the quark sector. Whether future experiments, such as at the LHC and (Super) B factories, will force us to modify this paradigm remains to be seen. We expect on theoretical grounds that there is New Physics, probably just around the corner, to solve the outstanding issue of the gauge hierarchy. The resolution of this problem together with the unification of the gauge couplings and the search of viable candidate(s) for dark matter requires a TeV scale New Physics. Assuming that supersymmetry is the most viable candidate for the impending New Physics to be discovered by experiments at the LHC, the central issue in the LHC era would be to pin down the underlying flavor aspects of this theory. However, if the current experimental trend is any indicator, then very likely the New Physics will be of the minimal flavor violating type, or something akin to it. Verifying this and, more importantly, quantifying the subdominant flavor structures in supersymmetry (or its alternative) is the next task to which theory and experiment have to gear themselves up.

Acknowledgments

I would like to congratulate Professor Riazuddin and the faculty of the National Centre for Physics in Islamabad on the auspicious occasion of the inauguration of the new campus. I have thoroughly enjoyed this symposium marking the inaugural and thank Professor Riazuddin and his staff for their kind hospitality. Thanks are also due to Alexander Parkhomenko for critically reading this manuscript and his helpful remarks.

References

1. N. Cabibbo, Phys. Rev. Lett. **10** (1963) 531.
2. M. Kobayashi and T. Maskawa, Prog. Theor. Phys. **49** (1973) 652.
3. E. Barberio, in Proc. of the XXIII International Symposium on Lepton and Photon Interactions at High Energy, Aug. 13-18, 2007, Daego, Korea. Below this is termed as the LP-2007 Symposium (http://chep.knu.ac.kr/LP07/).
4. D. Brown, in Proc. of the LP-2007 Symposium.
5. M. Nakao, in Proc. of the LP-2007 Symposium.
6. J. Wicht [BELLE Collaboration], arXiv:0711.0271 [hep-ex].
7. Heavy Flavor Averaging Group [HFAG]. The most recent HFAG averages can be seen in the eprint: E. Barberio et al. [arXiv:0704.3575 (hep-ex)] and from the HFAG website (http://www.slac.stanford.edu/xorg/hfag/).
8. V. M. Abazov *et al.* [D0 Collaboration], Phys. Rev. Lett. **97**, 021802 (2006) [arXiv:hep-ex/0603029].
9. A. Abulencia *et al.* [CDF - Run II Collaboration], Phys. Rev. Lett. **97**, 062003 (2006) [arXiv:hep-ex/0606027].
10. M. Morello [CDF Collaboration], Nucl. Phys. Proc. Suppl. **170**, 39 (2007) [arXiv:hep-ex/0612018].
11. A. Abulencia *et al.* [CDF Collaboration], Phys. Rev. Lett. **97**, 211802 (2006) [arXiv:hep-ex/0607021].
12. For the LHCb detector details and the physics potential, see the LHCb website: http://lhcb.cern.ch/lhcb/.
13. W. M. Yao *et al.* [Particle Data Group], J. Phys. G **33**, 1 (2006).
14. See, for example, the proceedings of the Workshop on Theory, Phenomenology and Experiments in Heavy Flavor Physics, Capri, Italy, 29-31 May 2006; Nucl. Phys. B, Proc. Suppl. 159 (2006) 213 p (Editors: G. Ricciardi and C. Sciacca).
15. L. Wolfenstein, Phys. Rev. Lett. **51**, 1945 (1983).
16. A. J. Buras, M. E. Lautenbacher and G. Ostermaier, Phys. Rev. D **50**, 3433 (1994) [hep-ph/9403384].
17. J. C. Hardy and I. S. Towner, Phys. Rev. Lett. **94**, 092502 (2005) [arXiv:nucl-th/0412050].

18. G. Savard *et al.*, Phys. Rev. Lett. **95**, 102501 (2005).
19. H. Leutwyler and M. Roos, Z. Phys. C **25**, 91 (1984).
20. For a recent review of Lattice QCD predictions for D-decays, see A. S. Kronfeld [Fermilab Lattice Collaboration], J. Phys. Conf. Ser. **46**, 147 (2006) [arXiv:hep-lat/0607011].
21. See, for example, A. V. Manohar and M. B. Wise, Cambridge Monogr. Part. Phys. Nucl. Phys. Cosmol. **10** (2000) 1.
22. J. Chay, H. Georgi, and B. Grinstein, Phys. Lett. B247 (1990) 399; M. Voloshin and M. Shifman, Sov. J. Nucl. Phys. 41 (1985) 120.
23. I.I. Bigi *et al.*, Phys. Lett. B293 (1992) 430; Phys. Lett B297 (1993) 477 (E); Phys. Rev. Lett. 71 (1993) 496.
24. A.V. Manohar and M.B. Wise, Phys. Rev. D49 (1994) 1310.
25. B. Blok, L. Koyrakh, M. A. Shifman and A. I. Vainshtein, Phys. Rev. D **49**, 3356 (1994); Phys. Rev. D **50**, 3572 (1994) (E) [hep-ph/9307247].
26. T. Mannel , Nucl. Phys. B **413**, 396 (1994) [hep-ph/9308262].
27. C. W. Bauer, Z. Ligeti, M. Luke, A. V. Manohar and M. Trott, Phys. Rev. D **70**, 094017 (2004) [hep-ph/0408002].
28. P. Gambino and N. Uraltsev, Eur. Phys. J. C **34**, 181 (2004) [hep-ph/0401063].
29. N. Isgur and M. B. Wise, Phys. Lett. B **232**, 113 (1989). Phys. Lett. B **237**, 527 (1990).
30. M. E. Luke, Phys. Lett. B **252**, 447 (1990).
31. J. D. Bjorken, SLAC-PUB-5278 *Talk given at Les Rencontre de la Valle d'Aoste, La Thuile, Italy, Mar 18-24, 1990.*
32. N. Uraltsev, Phys. Lett. B **501**, 86 (2001) [hep-ph/0011124]; J. Phys. G **27**, 1081 (2001) [hep-ph/0012336].
33. F. Jugeau, A. Le Yaouanc, L. Oliver and J. C. Raynal, Phys. Rev. D **71**, 054031 (2005) [arXiv:hep-ph/0407176].
34. J. P. Archambault and A. Czarnecki, Phys. Rev. D **70**, 074016 (2004) [hep-ph/0408021].
35. S. Hashimoto, A. S. Kronfeld, P. B. Mackenzie, S. M. Ryan and J. N. Simone, Phys. Rev. D **66**, 014503 (2002) [hep-ph/0110253].
36. M. Luke, eConf **C0304052**, WG107 (2003) [hep-ph/0307378].
37. M. Neubert, Phys. Rev. D49 (1994) 3392; D49 (1994) 4623; T. Mannel and M. Neubert, Phys. Rev. D50 (1994) 2037; I.I. Bigi *et al.*, Int. J. Mod. Phys. A9 (1994) 2467.
38. A. K. Leibovich, I. Low and I. Z. Rothstein, Phys. Rev. D **61**, 053006 (2000) [hep-ph/9909404]; Phys. Lett. B **486**, 86 (2000) [hep-ph/0005124]; Phys. Lett. B **513**, 83 (2001) [hep-ph/0105066].
39. M. Neubert, Phys. Lett. B **513**, 88 (2001) [hep-ph/0104280].
40. C. W. Bauer, M. E. Luke and T. Mannel, Phys. Rev. D **68**, 094001 (2003) [hep-ph/0102089].
41. G. P. Korchemsky and G. Sterman, Phys. Lett. B **340**, 96 (1994) [hep-ph/9407344].
42. C. W. Bauer, Z. Ligeti and M. E. Luke, Phys. Rev. D **64**, 113004 (2001) [arXiv:hep-ph/0107074].

43. C. W. Bauer, S. Fleming and M. E. Luke, Phys. Rev. D **63**, 014006 (2001) [arXiv:hep-ph/0005275].
44. C. W. Bauer, S. Fleming, D. Pirjol and I. W. Stewart, Phys. Rev. D **63**, 114020 (2001) [arXiv:hep-ph/0011336].
45. C. W. Bauer, D. Pirjol and I. W. Stewart, Phys. Rev. D **65**, 054022 (2002) [arXiv:hep-ph/0109045].
46. M. Beneke, A. P. Chapovsky, M. Diehl and T. Feldmann, Nucl. Phys. B **643**, 431 (2002) [arXiv:hep-ph/0206152].
47. B. O. Lange, M. Neubert and G. Paz, Phys. Rev. D **72**, 073006 (2005) [arXiv:hep-ph/0504071].
48. J. R. Andersen and E. Gardi, JHEP **0601**, 097 (2006) [arXiv:hep-ph/0509360].
49. P. Ball and R. Zwicky, Phys. Rev. D **71**, 014015 (2005) [arXiv:hep-ph/0406232].
50. E. Dalgic, A. Gray, M. Wingate, C. T. H. Davies, G. P. Lepage and J. Shigemitsu, Phys. Rev. D **73**, 074502 (2006) [Erratum-ibid. D **75**, 119906 (2007)] [arXiv:hep-lat/0601021].
51. M. Okamoto *et al.*, Nucl. Phys. Proc. Suppl. **140**, 461 (2005) [arXiv:hep-lat/0409116].
52. A. Abada, D. Becirevic, P. Boucaud, J. P. Leroy, V. Lubicz and F. Mescia, Nucl. Phys. B **619**, 565 (2001) [arXiv:hep-lat/0011065].
53. B. Grinstein and D. Pirjol, Phys. Rev. D **70**, 114005 (2004) [arXiv:hep-ph/0404250].
54. D. E. Acosta *et al.* [CDF Collaboration], Phys. Rev. Lett. **95**, 102002 (2005) [arXiv:hep-ex/0505091].
55. J. A. Aguilar-Saavedra *et al.* [ECFA/DESY LC Physics Working Group], arXiv:hep-ph/0106315.
56. A. J. Buras, W. Slominski and H. Steger, Nucl. Phys. B **245**, 369 (1984).
57. A. J. Buras, M. Jamin and P. H. Weisz, Nucl. Phys. B **347**, 491 (1990).
58. T. Inami and C. S. Lim, Prog. Theor. Phys. **65**, 297 (1981) [Erratum-ibid. **65**, 1772 (1981)].
59. M. Jamin and B. O. Lange, Phys. Rev. D **65**, 056005 (2002) [hep-ph/0108135].
60. A. A. Penin and M. Steinhauser, Phys. Rev. D **65**, 054006 (2002) [hep-ph/0108110].
61. J. G. Korner, A. I. Onishchenko, A. A. Petrov and A. A. Pivovarov, Phys. Rev. Lett. **91**, 192002 (2003) [arXiv:hep-ph/0306032].
62. S. Aoki *et al.* [JLQCD Collaboration], Phys. Rev. Lett. **91**, 212001 (2003) [arXiv:hep-ph/0307039].
63. A. Gray *et al.* [HPQCD Collaboration], Phys. Rev. Lett. **95**, 212001 (2005) [arXiv:hep-lat/0507015].
64. M. Okamoto, PoS **LAT2005**, 013 (2006) [arXiv:hep-lat/0510113].
65. E. Dalgic *et al.*, Phys. Rev. D **76**, 011501 (2007) [arXiv:hep-lat/0610104].
66. M. Bona *et al.* [UTfit Collaboration], arXiv:0707.0636 [hep-ph].
67. J. Charles *et al.* [CKMfitter Group], Eur. Phys. J. C **41**, 1 (2005) [arXiv:hep-ph/0406184]. Updates are taken from the CKMFitter Working Group URL:

http://ckmfitter.in2p3.fr.
68. M. Ciuchini and L. Silvestrini, Phys. Rev. Lett. **97**, 021803 (2006) [arXiv:hep-ph/0603114].
69. M. Endo and S. Mishima, Phys. Lett. B **640**, 205 (2006) [arXiv:hep-ph/0603251].
70. Z. Ligeti, M. Papucci and G. Perez, Phys. Rev. Lett. **97**, 101801 (2006) [arXiv:hep-ph/0604112].
71. A. Lenz and U. Nierste, JHEP **0706**, 072 (2007) [arXiv:hep-ph/0612167].
72. A. Ali and D. London, Eur. Phys. J. C **18**, 665 (2001) [arXiv:hep-ph/0012155].
73. A. Ali and D. London, Eur. Phys. J. C **9**, 687 (1999) [arXiv:hep-ph/9903535]; A. J. Buras, P. Gambino, M. Gorbahn, S. Jager and L. Silvestrini, Phys. Lett. B **500**, 161 (2001) [arXiv:hep-ph/0007085]; G. D'Ambrosio, G. F. Giudice, G. Isidori and A. Strumia, Nucl. Phys. B **645**, 155 (2002) [arXiv:hep-ph/0207036].
74. A. Ali and Z. Z. Aydin, Nucl. Phys. B **148**, 165 (1979).
75. G. Buchalla, A. J. Buras and M. E. Lautenbacher, Rev. Mod. Phys. **68**, 1125 (1996) [arXiv:hep-ph/9512380].
76. T. Hurth, Rev. Mod. Phys. **75**, 1159 (2003) [arXiv:hep-ph/0212304].
77. C. Bobeth, M. Misiak and J. Urban, Nucl. Phys. B **574**, 291 (2000) [hep-ph/9910220].
78. M. Misiak and M. Steinhauser, Nucl. Phys. B **683**, 277 (2004) [hep-ph/0401041].
79. K. G. Chetyrkin, M. Misiak and M. Münz, Phys. Lett. B **400**, 206 (1997) [Erratum-ibid. B **425**, 414 (1998)] [hep-ph/9612313].
80. M. Gorbahn and U. Haisch, Nucl. Phys. B **713**, 291 (2005) [arXiv:hep-ph/0411071].
81. M. Gorbahn, U. Haisch and M. Misiak, Phys. Rev. Lett. **95**, 102004 (2005) [arXiv:hep-ph/0504194].
82. M. Czakon, U. Haisch and M. Misiak, JHEP **0703**, 008 (2007) [arXiv:hep-ph/0612329].
83. C. Greub, T. Hurth and D. Wyler, Phys. Rev. D **54**, 3350 (1996) [hep-ph/9603404].
84. I. Blokland, A. Czarnecki, M. Misiak, M. Slusarczyk and F. Tkachov, Phys. Rev. D **72**, 033014 (2005) [arXiv:hep-ph/0506055].
85. H. M. Asatrian, A. Hovhannisyan, V. Poghosyan, T. Ewerth, C. Greub and T. Hurth, Nucl. Phys. B **749**, 325 (2006) [arXiv:hep-ph/0605009].
86. H.M. Asatrian, T. Ewerth, A. Ferroglia, P. Gambino, C. Greub and G. Ossola, in preparation.
87. A. J. Buras, A. Czarnecki, M. Misiak and J. Urban, Nucl. Phys. B **611**, 488 (2001) [arXiv:hep-ph/0105160].
88. A. J. Buras, A. Czarnecki, M. Misiak and J. Urban, Nucl. Phys. B **631**, 219 (2002) [arXiv:hep-ph/0203135].
89. K. Bieri, C. Greub and M. Steinhauser, Phys. Rev. D **67**, 114019 (2003) [arXiv:hep-ph/0302051].
90. M. Misiak and M. Steinhauser, Nucl. Phys. B **764**, 62 (2007) [arXiv:hep-

ph/0609241].
91. M. Misiak *et al.*, Phys. Rev. Lett. **98**, 022002 (2007) [arXiv:hep-ph/0609232].
92. A. Ali and C. Greub, Z. Phys. C **49**, 431 (1991); Phys. Lett. B **361**, 146 (1995) [hep-ph/9506374]; N. Pott, Phys. Rev. D **54**, 938 (1996) [hep-ph/9512252].
93. K. Melnikov and A. Mitov, Phys. Lett. B **620**, 69 (2005) [arXiv:hep-ph/0505097].
94. J. R. Ellis, S. Heinemeyer, K. A. Olive, A. M. Weber and G. Weiglein, JHEP **0708**, 083 (2007) [arXiv:0706.0652 [hep-ph]].
95. T. Becher and M. Neubert, Phys. Rev. Lett. **98**, 022003 (2007) [arXiv:hep-ph/0610067].
96. B. Grinstein, M. J. Savage and M. B. Wise, Nucl. Phys. B **319**, 271 (1989).
97. F. Kruger and L. M. Sehgal, Phys. Lett. B **380**, 199 (1996) [arXiv:hep-ph/9603237].
98. A. F. Falk, M. E. Luke and M. J. Savage, Phys. Rev. D **49**, 3367 (1994) [arXiv:hep-ph/9308288].
99. A. Ali, G. Hiller, L. T. Handoko and T. Morozumi, Phys. Rev. D **55**, 4105 (1997) [arXiv:hep-ph/9609449].
100. C. W. Bauer and C. N. Burrell, Phys. Rev. D **62**, 114028 (2000) [arXiv:hep-ph/9911404]; Phys. Lett. B **469**, 248 (1999) [arXiv:hep-ph/9907517].
101. J. W. Chen, G. Rupak and M. J. Savage, Phys. Lett. B **410**, 285 (1997) [arXiv:hep-ph/9705219]; G. Buchalla, G. Isidori and S. J. Rey, Nucl. Phys. B **511**, 594 (1998) [arXiv:hep-ph/9705253].
102. A. Ali, E. Lunghi, C. Greub and G. Hiller, Phys. Rev. D **66**, 034002 (2002) [hep-ph/0112300].
103. H. H. Asatryan, H. M. Asatrian, C. Greub and M. Walker, Phys. Rev. D **65**, 074004 (2002) [hep-ph/0109140]. H. H. Asatrian, H. M. Asatrian, C. Greub and M. Walker, Phys. Lett. B **507**, 162 (2001) [hep-ph/0103087].
104. A. Ghinculov, T. Hurth, G. Isidori and Y. P. Yao, Nucl. Phys. B **685**, 351 (2004) [hep-ph/0312128].
105. C. Bobeth, P. Gambino, M. Gorbahn and U. Haisch, JHEP **0404**, 071 (2004) [arXiv:hep-ph/0312090].
106. A. Ali and G. Hiller, Phys. Rev. D **60**, 034017 (1999) [arXiv:hep-ph/9807418].
107. K. S. M. Lee, Z. Ligeti, I. W. Stewart and F. J. Tackmann, Phys. Rev. D **74**, 011501 (2006) [arXiv:hep-ph/0512191].
108. Z. Ligeti and F. J. Tackmann, Phys. Lett. B **653**, 404 (2007) [arXiv:0707.1694 [hep-ph]].
109. A. Ali, G. F. Giudice and T. Mannel, Z. Phys. C **67**, 417 (1995) [hep-ph/9408213].
110. G. Hiller and F. Krüger, Phys. Rev. D **69**, 074020 (2004) [hep-ph/0310219].
111. P. Gambino, U. Haisch and M. Misiak, Phys. Rev. Lett. **94**, 061803 (2005) [arXiv:hep-ph/0410155].
112. A. Ali, T. Mannel and T. Morozumi, Phys. Lett. B **273**, 505 (1991).
113. A. Ghinculov, T. Hurth, G. Isidori and Y. P. Yao, Nucl. Phys. B **648**, 254 (2003) [hep-ph/0208088].
114. H. M. Asatrian, K. Bieri, C. Greub and A. Hovhannisyan, Phys. Rev. D **66**,

094013 (2002) [hep-ph/0209006].
115. G. Burdman, Phys. Rev. D **57**, 4254 (1998) [hep-ph/9710550].
116. A. Ali, P. Ball, L. T. Handoko and G. Hiller, Phys. Rev. D **61**, 074024 (2000) [hep-ph/9910221].
117. M. Beneke and T. Feldmann, Nucl. Phys. B **592**, 3 (2001) [hep-ph/0008255].
118. M. Beneke, T. Feldmann and D. Seidel, Nucl. Phys. B **612** (2001) 25 [hep-ph/0106067].
119. A. Ali, G. Kramer and G. h. Zhu, Eur. Phys. J. C **47**, 625 (2006) [arXiv:hep-ph/0601034].
120. M. Beneke, G. Buchalla, M. Neubert and C. T. Sachrajda, Phys. Rev. Lett. **83**, 1914 (1999) [arXiv:hep-ph/9905312].
121. M. Beneke, G. Buchalla, M. Neubert and C. T. Sachrajda, Nucl. Phys. B **606**, 245 (2001) [arXiv:hep-ph/0104110].
122. A. Ishikawa *et al.*, Phys. Rev. Lett. **96**, 251801 (2006) [arXiv:hep-ex/0603018].
123. Y. Y. Keum, H. n. Li and A. I. Sanda, Phys. Lett. B **504**, 6 (2001) [hep-ph/0004004]; Y. Y. Keum and A. I. Sanda, Phys. Rev. D **67**, 054009 (2003) [hep-ph/0209014].
124. A. Ali and A. Y. Parkhomenko, Eur. Phys. J. C **23**, 89 (2002) [arXiv:hep-ph/0105302].
125. S. W. Bosch and G. Buchalla, Nucl. Phys. B **621**, 459 (2002) [arXiv:hep-ph/0106081].
126. A. L. Kagan and M. Neubert, Phys. Lett. B **539**, 227 (2002) [hep-ph/0110078].
127. A. Ali, E. Lunghi and A. Y. Parkhomenko, Phys. Lett. B **595**, 323 (2004) [arXiv:hep-ph/0405075].
128. S. W. Bosch and G. Buchalla, JHEP **0501**, 035 (2005) [arXiv:hep-ph/0408231].
129. M. Beneke, T. Feldmann and D. Seidel, Eur. Phys. J. C **41**, 173 (2005) [arXiv:hep-ph/0412400].
130. A. Ali and A. Parkhomenko, arXiv:hep-ph/0610149.
131. P. Ball and R. Zwicky, JHEP **0604**, 046 (2006) [arXiv:hep-ph/0603232].
132. Y. Y. Keum, M. Matsumori and A. I. Sanda, Phys. Rev. D **72**, 014013 (2005) [arXiv:hep-ph/0406055].
133. C. D. Lu, M. Matsumori, A. I. Sanda and M. Z. Yang, Phys. Rev. D **72**, 094005 (2005) [Erratum-ibid. D **73**, 039902 (2006)] [arXiv:hep-ph/0508300].
134. J. g. Chay and C. Kim, Phys. Rev. D **68**, 034013 (2003) [arXiv:hep-ph/0305033].
135. T. Becher, R. J. Hill and M. Neubert, Phys. Rev. D **72**, 094017 (2005) [arXiv:hep-ph/0503263].
136. M. Beneke and T. Feldmann, Nucl. Phys. B **685**, 249 (2004) [arXiv:hep-ph/0311335].
137. B. O. Lange and M. Neubert, Nucl. Phys. B **690**, 249 (2004) [Erratum-ibid. B **723**, 201 (2005)] [arXiv:hep-ph/0311345].
138. M. Beneke, Y. Kiyo and D. s. Yang, Nucl. Phys. B **692**, 232 (2004) [arXiv:hep-ph/0402241].

139. M. Beneke and D. Yang, Nucl. Phys. B **736**, 34 (2006) [arXiv:hep-ph/0508250].
140. R. J. Hill, T. Becher, S. J. Lee and M. Neubert, JHEP **0407**, 081 (2004) [arXiv:hep-ph/0404217].
141. T. Becher and R. J. Hill, JHEP **0410**, 055 (2004) [arXiv:hep-ph/0408344].
142. A. Ali, B. D. Pecjak and C. Greub, arXiv:0709.4422 [hep-ph].
143. A. Ali and V. M. Braun, Phys. Lett. B **359**, 223 (1995) [hep-ph/9506248].
144. Y. Li, C. D. Lu and W. Wang, arXiv:0711.0497 [hep-ph].
145. K. Koneke [Babar Collaboration], arXiv:0710.4855 [hep-ex].
146. M. Gronau and D. London, Phys. Rev. Lett. **65**, 3381 (1990).
147. A. J. Buras and R. Fleischer, Phys. Lett. B **341**, 379 (1995) [hep-ph/9409244].
148. N. G. Deshpande and X. G. He, Phys. Rev. Lett. **74** (1995) 26 [Erratum-ibid. **74** (1995) 4099] [hep-ph/9408404]. M. Gronau, O. F. Hernandez, D. London and J. L. Rosner, Phys. Rev. D **52**, 6374 (1995) [hep-ph/9504327].
149. M. Gronau and J. L. Rosner, Phys. Rev. D **66**, 053003 (2002) [Erratum-ibid. D **66**, 119901 (2002)] [hep-ph/0205323].
150. M. Ciuchini, E. Franco, G. Martinelli, M. Pierini and L. Silvestrini, Phys. Lett. B **515**, 33 (2001) [hep-ph/0104126].
151. A. J. Buras and L. Silvestrini, Nucl. Phys. B **569**, 3 (2000) [hep-ph/9812392].
152. A. B. Carter and A. I. Sanda, Phys. Rev. D **23**, 1567 (1981). I. I. Y. Bigi and A. I. Sanda, Nucl. Phys. B **193**, 85 (1981).
153. A. Ali, E. Lunghi and A. Y. Parkhomenko, Eur. Phys. J. C **36**, 183 (2004). [hep-ph/0403275].
154. M. Ciuchini, E. Franco, G. Martinelli, M. Pierini and L. Silvestrini, *Prepared for 32nd International Conference on High-Energy Physics (ICHEP 04), Beijing, China, 16-22 Aug 2004.*
155. C. W. Bauer, D. Pirjol, I. Z. Rothstein and I. W. Stewart, Phys. Rev. D **70**, 054015 (2004) [hep-ph/0401188].
156. C. W. Bauer, I. Z. Rothstein and I. W. Stewart, Phys. Rev. D **74**, 034010 (2006) [arXiv:hep-ph/0510241].
157. M. Gronau and D. Wyler, Phys. Lett. B **265**, 172 (1991).
158. M. Gronau and D. London., Phys. Lett. B **253**, 483 (1991).
159. D. Atwood, I. Dunietz and A. Soni, Phys. Rev. Lett. **78**, 3257 (1997) [hep-ph/9612433].
160. D. Atwood, I. Dunietz and A. Soni, Phys. Rev. D **63**, 036005 (2001) [hep-ph/0008090].
161. M. Gronau, Phys. Lett. B **557**, 198 (2003) [hep-ph/0211282].
162. D. Atwood and A. Soni, Phys. Rev. D **71**, 013007 (2005) [arXiv:hep-ph/0312100].
163. A. Giri, Y. Grossman, A. Soffer and J. Zupan, Phys. Rev. D **68**, 054018 (2003) [hep-ph/0303187].
164. A. J. Buras, R. Fleischer, S. Recksiegel and F. Schwab, Acta Phys. Polon. B **36**, 2015 (2005) [arXiv:hep-ph/0410407]; Nucl. Phys. B **697**, 133 (2004) [hep-ph/0402112]; Phys. Rev. Lett. **92**, 101804 (2004) [hep-ph/0312259].
165. R. Fleischer and T. Mannel, Phys. Rev. D **57**, 2752 (1998) [hep-ph/9704423].

166. M. Neubert and J. L. Rosner, Phys. Rev. Lett. **81**, 5076 (1998) [hep-ph/9809311]; Phys. Lett. B **441**, 403 (1998) [hep-ph/9808493].
167. H. J. Lipkin, Phys. Lett. B **445**, 403 (1999) [hep-ph/9810351].
168. A. J. Buras and R. Fleischer, Eur. Phys. J. C **16**, 97 (2000) [hep-ph/0003323]; Eur. Phys. J. C **11**, 93 (1999) [hep-ph/9810260].
169. T. Yoshikawa, Phys. Rev. D **68**, 054023 (2003) [arXiv:hep-ph/0306147]; J. Korean Phys. Soc. **45**, S479 (2004) [arXiv:hep-ph/0312097].
170. M. Beneke and M. Neubert, Nucl. Phys. B **675**, 333 (2003). [hep-ph/0308039].
171. R. Fleischer and S. Recksiegel, Eur. Phys. J. C **38**, 251 (2004) [arXiv:hep-ph/0408016].
172. M. Beneke, G. Buchalla, C. Greub, A. Lenz and U. Nierste, Phys. Lett. B **459**, 631 (1999) [arXiv:hep-ph/9808385].
173. V. M. Abazov *et al.* [D0 Collaboration], Phys. Rev. Lett. **98**, 121801 (2007) [arXiv:hep-ex/0701012].
174. T. Kuhr [CDF Collaboration], arXiv:0710.1789 [hep-ex].
175. For a recent review of CP violation at the LHC, see R. Fleischer, *In the Proceedings of International Conference on Heavy Quarks and Leptons (HQL 06), Munich, Germany, 16-20 Oct 2006, pp 020* [arXiv:hep-ph/0703112].
176. O. Schneider, LHCb Upgrade Workshop, Edinburgh, January 11 -12, 2007.
177. P. Mack [CDF Collaboration], arXiv:0710.2502 [hep-ex].
178. D0 Public Note 5344-CONF based on 2 fb^{-1} data, submitted to the EPS Conference, 2007.
179. G. Buchalla and A. J. Buras, Nucl. Phys. B **400**, 225 (1993); *ibid.* **548**, 309 (1999) [hep-ph/9901288]; M. Misiak and J. Urban, Phys. Lett. B **451**, 161 (1999) [hep-ph/9901278]; A. J. Buras, Phys. Lett. B **566**, 115 (2003) [hep-ph/0303060].
180. R. Dermisek, S. Raby, L. Roszkowski and R. Ruiz De Austri, JHEP **0304**, 037 (2003) [hep-ph/0304101].
181. J. R. Ellis, S. Heinemeyer, K. A. Olive and G. Weiglein, JHEP **0502**, 013 (2005) [arXiv:hep-ph/0411216].
182. H. J. Lipkin, Phys. Lett. B **621**, 126 (2005) [arXiv:hep-ph/0503022].
183. M. Beneke, J. Rohrer and D. Yang, Nucl. Phys. B**774**, 64 (2007) [hep-ph/0612290]; M. Beneke and M. Neubert, Nucl. Phys. B**675**, 333(2003) [hep-ph/0308039].
184. A. Williamson and J. Zupan, Phys. Rev. D**74**, 014003(2006); Erratum-ibid. D**74**, 039901 (2006) [hep-ph/0601214].
185. A. Ali, G. Kramer, Y. Li, C. D. Lu, Y. L. Shen, W. Wang and Y. M. Wang, Phys. Rev. D **76**, 074018 (2007) [arXiv:hep-ph/0703162].

Quantum gravity and black holes

Viqar Husain

Department of Mathematics and Statistics, University of New Brunswick, Fredericton, NB E3B 2S9, Canada.
E-mail: vhusain@unb.ca

We give an overview of some of the ideas involved in formulating a quantum theory of gravity, such as fundamental discreteness of spacetime and background independent quantization We describe the role played by these ingredients in understanding gravitational collapse in quantum gravity.

Keywords: Quantum gravity, gravitational collapse, black holes.

1. Classical setting

In non-gravitational theories the spacetime metric is not a dynamical variable. It is a fixed kinematical structure that forms a part of the stage, or background, on which dynamics takes place. A fixed metric appears in at least the kinetic energy part of the lagrangian of any conventional physical system, whether it is relativistic or not. In Newtonian mechanics it is the (often invisible) Euclidean metric q_{ab} that appears in the particle action

$$S = \int \left(\; q_{ab}\dot{x}^a\dot{x}^b - V(x)\right) \; dt, \tag{1}$$

whereas in special relativistic field theory it is the Minkowski metric η^{ab}. For example in the lagrangian of the electromagnetic field, the vector potential A_a is a 1-form, so a metric is required to write down the action

$$S_{em} = \frac{1}{4}\int \sqrt{-\eta} F_{ab}F_{cd}\eta^{ac}\eta^{db} \; d^4x. \tag{2}$$

More generally these actions may be written with any invertible metric, giving theories on arbitrary backgrounds[1,2]. It is the symmetries of the metric that are responsible for the usual conservations laws. In particular the notion of special classes of reference frames, such as inertial ones, are concomitant with spacetime symmetries. For metrics that do not have any symmetries, there is no special class of reference frames, no preferred form of a Hamiltonian, and therefore no notion of the invariance of physics as seen by observers in relative motion. For example replacing the Euclidean or Minkowski metrics in (1) or (2) with an arbitrary time dependent metric leads to time dependent kinetic energies and propagators. The simplest example of this is provided by a free particle Hamiltonian on a time dependent background prescribed by a metric $g_{ab}(t)$:

$$H = \frac{1}{2m}\, g^{ab}(t) p_a p_b. \tag{3}$$

More generally, for quantum field theory on general backgrounds, there is also no invariant notion of "particle," no global Poincare invariance, and therefore no preferred vacuum state.

In theories where gravitation is described by a dynamical metric field these issues are further magnified because the metric itself is apriori unknown. It is varied in an action, and so arises only as a solution of a set of equations of motion with coupling to matter degrees of freedom.

Due to the special role played by the metric in gravitational theory, all known physical theories can be divided into two types: those that utilize a fixed background metric, and those that do not. These are referred to as background dependent and independent theories – the latter being the ones in which the metric is varied in an action to obtain its equations of motion. General relativity and its variations are therefore background independent classical theories.

1.1. *Black hole and cosmological solutions*

For quantum theories of matter on a fixed background metric, three solutions of Einstein's equations have played an important role historically. These are Minkowski space for gravitational waves, back hole solutions for understanding the effect of horizons on quantum fields[1,2], and cosmological solutions for understanding structure formation[3]. From the theoretical perspective, each of these in their simplest realizations have provided questions and approaches for quantum gravity that are still actively studied.

The simplest black hole is described by the Schwarzschild metric

$$ds^2 = -\left(1 - \frac{2M}{r}\right) dt^2 + \left(1 - \frac{2M}{r}\right)^{-1} dr^2 + r^2 d\Omega^2 \tag{4}$$

where M is the mass of the black hole and $d\Omega^2$ is the unit 2-sphere metric. This is a stable and singular (at $r = 0$) vacuum solution that arises as the end point of classical gravitational collapse for certain types of initial conditions[4,5]. The surface $r = 2M$ is a one way barrier called the event horizon – no signal can escape from a radial location $r < 2M$ to one where $r > 2M$. Specially because of the horizon, this metric provides a very interesting background on which to study the behaviour of a quantum field. This is the starting point that has led to the puzzles of black hole physics, such as Hawking radiation[6] and the associated information loss problem.

For cosmological applications perhaps the most important solution physically is the homogeneous and isotropic Friedmann-Robertson-Walker metric. The form of the solution where 3-space is flat is

$$ds^2 = -dt^2 + a^2(t)(dx^2 + dy^2 + dz^2). \tag{5}$$

The sole function in the metric is determined by the matter content on the r.h.s. of Einstein's equation, and is usually taken to be a fluid with energy density $\rho(t)$ and pressure $P(t)$ satisfying an equation of state $P = k\rho$. This solution is singular when the scale factor $a(t)$ goes to zero. This is the big bang curvature singularity, where the matter energy density also diverges.

Both these solutions present important questions for potential theories of quantum gravity. Since curvature and energy density divergences are believed to be unphysical, are these singularities resolved by quantization similar to the Coulomb potential singularity in atomic theory? If metric functions are quantized like any other field, what is to be replaced by the sharply defined classical notion of event horizon? What is a black hole in quantum gravity? How do these metrics "emerge" at large distance scales from a quantum theory? What are the quantum gravity corrections to these metrics? At present there are no compelling answers to these questions in any approach to quantum gravity.

2. Quantum theory

As we have seen, all conventional theories are background dependent. The usual approach to quantization makes essential use of a fixed metric. This is evident in the Hilbert space inner-product used for computing expectation values. For example, for the quantum mechanics of a particle moving on a curved space given by a 3-metric q_{ab}, we have

$$\langle\phi|\psi\rangle = \int d^3x\sqrt{q}\ \phi^*(x)\psi(x), \tag{6}$$

where the determinant factor $\sqrt{q}$ is usually invisible because it is unity for flat space in Cartesian coordinates. This factor is necessary because the integrand must be a scalar density. Similarly, in standard quantum field theory the Minkowski metric and its symmetries play a fundamental role as we have already noted. This poses the first challenge for quantum gravity: How do we quantize if the metric itself is to be a quantum variable?

2.1. *Semiclassical approximation*

A hybrid approach called the semi-classical approximation was introduced in the 1970s[1,2]. It bypasses the issue of a quantum metric by treating it as a classical field and matter as a quantum field. The postulated modification of Einstein's equation is

$$G_{ab} = 8\pi G \ \langle\psi|\hat{T}_{ab}|\psi\rangle, \tag{7}$$

where ψ is a prescribed quantum state, usually taken to be a Fock vacuum for some observer, but it can be anything. Given such a state, the problem is to calculate the semiclassical metric using this equation. The resulting metric is dependent on the choice of matter quantum state, so we can write its solutions as $g^{sc}_{ab}(x;|\psi\rangle)$. (If we seek a perturbative solution in the form $g^{sc} = g^c + \delta g$ and keep terms to 0th. order in the quantum correction to the metric δg, we recover the fixed background approximation.)

This is clearly quite different from the classical equations, where both the matter and metric fields are solved for together. Since the metric appears on both sides of the equations, it raises both consistency and interpretational questions. For instance, for what classes of metrics can one find interesting solutions, and what is the interpretation of the metric if the state $|\psi\rangle$ is a linear combination? Furthermore, the expectation value of the stress-energy tensor is divergent and requires a regularization and renormalization. Do semiclassical metrics depend on this?

There is an analogous quantum mechanics setting that illustrates the rather odd nature of the semiclassical approximation. Consider the Schrodinger equation

$$i\hbar\frac{\partial\psi}{\partial t} = -\frac{\hbar^2}{2m}q^{ab}(t)\ \frac{\partial^2\psi}{\partial x^a\partial x^b} + V(x)\psi \tag{8}$$

as an equation for the 3-space metric q_{ab} given a quantum state ψ. The question of self consistency here is immediate because of the forgoing discussion on the metric dependance of Schrodinger quantization – we need a metric to define the inner product, and this is usually the same as the metric appearing in the kinetic energy term.

Despite these issues the approximation has been useful for computing the rate of mass loss of black a hole due to Hawking radiation. This uses a family of Schwarzschild metrics with time dependent mass $M(t)$. Up to a constant of order unity, the rate of mass loss can be derived by simply using dimensional analysis, with the sole input that a black hole of mass M radiates like a black body at a temperature T proportional to $1/M$. The Stefan-Boltzmann law states that the energy radiated per unit area per unit time from a black body is proportional to $T^4 \sim M^{-4}$. So the rate of mass loss is proportional to $4\pi R^2 T^4$ where $R = 2M$ is the black hole radius. Putting these together we have

$$\frac{dM}{dt} = -\alpha \, \frac{M_P}{t_P} \left(\frac{M_P}{M} \right)^2 , \tag{9}$$

where α is a dimensionless constant and M_P and t_P are the Planck mass and time. This formula breaks down as $M \to 0$, where quantum gravitational considerations must come into play. We shall see below how incorporating a particular realization of fundamental discreteness modifies this formula.

Although the semiclassical approximation incorporates all three constants c, $\hbar$ and G, its scope is limited and it cannot be considered even close to fundamental. In addition to black hole physics, it is also used in cosmology where it forms the basis for a theory of structure formation. In both contexts quantum gravity is needed to provide an understanding of how to go beyond the approximation.

2.2. *Background independent quantization*

We have seen that if the metric is viewed as a dynamical field to be quantized, there is a need for a quantization scheme that does not require a metric. There are in fact many familiar examples of such theories. Consider for instance a spin system such as the Ising model with spin 1/2 at each lattice point. Such models are defined on a specified discrete space with a length scale and a notion of distance between lattice points, so a fixed metric is implicit. However it is important to note that the metric plays no role in defining the inner product on states. For example, for a lattice of N sites, we have two states per site so the space of states may be written as $|s, n\rangle$, where $s = \pm 1$ and $n = 1, 2 \cdots N$, with the inner product

$$\langle s', m | s, n \rangle = \delta_{s,s'} \delta_{m,n} . \tag{10}$$

This example suggests that space discreteness and background independent quantization may be linked, since the inner product is not given by a continuum space integral, but by a sum in an internal spin space. [a]

A similar situation arises if we wish to quantize a spinless particle system on a lattice. It is apparent that on any lattice we cannot have the usual momentum operator $\hat{p}_x = i\hbar\partial_x$ because there can be no generator of infinitesimal translations without space modeled as a continuum. Therefore only translations operators that provide hopping between lattice sites can be realized in quantum theory. So we are forced to consider obtaining a quantum theory not from the canonical Poisson bracket $\{x,p\} = 1$ but from

$$\{x, e^{ip\lambda}\} = i\lambda e^{ip\lambda}. \tag{11}$$

Consider the specific one -dimensional lattice with points $x_n = n\lambda$ ($n = \cdots -1, 0, 1\cdots$). As for an Ising spin system, we can use a Hilbert space spanned by the basis states $|n\rangle$ with $\langle m|n\rangle = \delta_{n,m}$, but now with spin $s = 0$. The position and translation operators are realized on this basis as

$$\hat{x}|n\rangle = \lambda|n\rangle, \qquad \hat{U}_\lambda|n\rangle \equiv \widehat{e^{i\lambda p}}|n\rangle = |n+1\rangle. \tag{12}$$

This is a λ dependent, but background independent quantization of the particle. [b] Any other operator of interest must be written using these basic operators. For example a Hamiltonian with potential $V(x) = -1/x$ that has a well defined action on the basis states is[8]

$$\hat{H} = \frac{1}{2m\lambda^2}(2 - \hat{U}_\lambda - \hat{U}_\lambda^\dagger) - \frac{1}{\lambda^2}\left(\hat{U}_\lambda\sqrt{|x|}\,\hat{U}_\lambda^\dagger - \hat{U}_\lambda^\dagger\sqrt{|x|}\,\hat{U}_\lambda\right)^2 \tag{13}$$

The action of the second term on a basis state is

$$-\frac{1}{\lambda^2}\left(\sqrt{|x_{n+1}|} - \sqrt{|x_{n-1}|}\right)^2, \tag{14}$$

which for large x behaves like $-1/x$, but is zero at $n = 0$. [c] This is the prototype of a singularity resolution method used in quantum gravity[10–13].

We summarize this section with two observations. If space is considered to be discrete at a fundamental level then (i) the translation and position operators are the natural building blocks of a background independent quantization, and (ii) variables such as $1/x$ that diverge classically can be

---

[a] An early use of a type of "spin" quantization in quantum gravity is in the strong coupling limit, $G_N \to \infty$[7].

[b] The discrete structure in loop quantum gravity[9] arises due to the use of a non-separable Hilbert space spanned by the spin network states. These states are supported on graphs embedded in a 3−manifold. This Hilbert space therefore utilizes all possible lattices.

[c] This expression is also the square of a finite difference approximation of the derivative $2\partial_x\sqrt{|x|}$, and the eigenvalue problem arising from the Hamiltonian (13) may be viewed as a Schrodinger equation discretized on a lattice.

naturally regulated. These ingredients generalize to field theory, where they provide an approach to studying black hole formation and evolution in a fully quantum setting.

3. Gravitational collapse in quantum gravity

One of the outstanding problems in theoretical physics is the incomplete understanding at the quantum level of the formation, and subsequent evolution of black holes in a quantum theory of gravity. Although a subject of study for over three decades, it is fair to say that, in spite of partial results in string theory and loop quantum gravity, there is no widely accepted answer to many of the puzzles of black hole physics. This is largely because *there has been no study of quantum dynamical collapse* in these approaches. Rather, progress has focused mainly on explanations of the microscopic origin of the entropy of *static* black holes from state counting. A spacetime picture of black hole formation from matter collapse, and its subsequent evolution is not available in any approach to quantum gravity at the present time.

Hawking's original derivation of black hole radiation uses the semiclassical approximation with spherically symmetric gravity minimally coupled to a massless scalar field. The full model is a non-linear 2d field theory describing the coupled system of the metric and scalar field degrees of freedom. Gravitational collapse in the classical theory in this model has been carefully studied numerically[4,5], but its full quantization has never been addressed. In this section we outline an approach to the problem that incorporates fundamental discreteness, singularity resolution, and a notion of quantum black hole is a dynamical setting.

3.1. *Scalar field model*

Our starting point is the Arnowitt-Deser-Misner (ADM) hamiltonian formulation of general relativity where the spatial part of the metric q_{ab} and its conjugate momentum π^{ab} are the phase variables. Coupled to a scalar field we have in addition the canonical variables (ϕ, P_ϕ). We are interested in the spherically symmetric reduction of this system, which is the one studied classically. This is a non-linear 2-dimensional field theory given by an Hamiltonian action of the form

$$S = \frac{1}{2G}\int dtdr\left(P_R\dot{R} + P_\Lambda\dot{\Lambda} + P_\phi\dot{\phi} - H\right) + \text{ surface terms,} \qquad (15)$$

where P_R and P_Λ are the momenta conjugate to the metric variables as they appear in the line element

$$ds^2 = \Lambda^2(r,t)dr^2 + R^2(r,t)d\Omega^2, \tag{16}$$

and the Hamiltonian H is a linear combination of constraints. Its detailed form is not necessary for our discussion here[14].

The central point concerning quantization of this field theory model is that it can be carried out on a radial lattice using field translation and configuration operators, in much the same way as for quantum mechanics on a lattice. To illustrate this let us focus on the metric variable $R(r,t)$ and its conjugate momentum $R(r,t)$, with the understanding that the same process applies to the other canonical variables. We seek to represent the Poisson bracket

$$\left\{R_f, e^{i\lambda P_R(r)}\right\} \equiv \left\{\int_0^\infty R(r,t)f(r)\; dr, e^{i\lambda P_R(r,t)}\right\} = i2G\lambda f(r)e^{i\lambda P_R(r)}, \tag{17}$$

on an arbitrary lattice. Here $f(r)$ is a smearing function with specific fall off conditions, and λ is a fixed constant that represents a field translation unit. For $f(r)$ we use a Gaussian peaked at lattice points.

To find a representation of this Poisson bracket consider a radial lattice of points r_i, $i = 1 \cdots N$, on which values of the field R given by $R_1 \cdots R_N$ are specified. We can write such a state as $|R_1 \ldots R_N\rangle$, with the inner product

$$\langle R_1 \ldots R_N | R_1' \ldots R_N' \rangle = \delta_{R_1,R_1'} \ldots \delta_{R_N,R_N'}.$$

The configuration and translation operators are realized on this basis by

$$\hat{R}_f\; |R_1 \ldots R_N\rangle = \sum_k R_k f(r_k) |R_1 \ldots R_N\rangle, \tag{18}$$

$$\widehat{e^{i\lambda P_R(r_j)}} |R_1 \ldots R_N\rangle = |R_1 \ldots, R_j - \lambda, \ldots R_N\rangle, \tag{19}$$

This is a generalization to field theory of the lattice quantum mechanics discussed above.

All operators of interest can now be constructed by appropriate sums of products of these basic ones, and their analogs for the other canonical variables. The basis of states extends in the obvious way for the other fields ϕ and Λ. Among these are the operators corresponding to negative powers of the field such as $1/R$, which plays a central role in discussions of singularity avoidance because curvatures are proportional to its powers. One way of writing such an operator is by using the identity

$$\frac{1}{|R_f|} = \left(\frac{2}{iGf(r)}\; e^{-iP_R(r)} \left\{\sqrt{|R_f|}, e^{iP_R(r)}\right\}\right)^2, \tag{20}$$

and replacing Poisson brackets by operators[10,13]. The spectrum of this operator and its variants are bounded. Thus inverse powers of the field $R(r,t)$ that appear in the classical evolution equations become finite in the corresponding quantum equations, (which are obtained by taking the expectation values of operator equations in suitable states). In this sense effective equations become singularity free.

3.2. *Quantum black hole formation*

Let us turn to defining what we mean by a black hole in a dynamical quantum theory. For static or stationary classical black holes, an essential ingredient is the event horizon. This is a global non-dynamical classical property of a spacetime ("the boundary of the past of future null infinity"), and appears unsuitable for characterizing a local quantum observable. There is however a more local observable that is used to detect black hole formation in classical simulations of gravitational collapse.[15] This involves computing the outward and inward null expansions $\theta_\pm$ associated with closed 2-surfaces embedded in a spatial 3−surface. These observables are functions of the phase space variables, and so can be converted to operators using the definitions of the basic operators given above. We can then define a quantum black hole state $|\Psi\rangle$ to be one that satisfies the conditions[16]

$$\langle\Psi|\hat{\theta}_+(r_k)|\Psi\rangle = 0, \quad \text{and} \quad \langle\Psi|\hat{\theta}_-(r_k)|\Psi\rangle < 0. \tag{21}$$

for some r_k. The corresponding *horizon size* is given by $R_H = \langle\Psi|\hat{R}(r_k)|\Psi\rangle$.

The ingredients presented here form a computational framework for studying gravitational collapse in quantum gravity. A first step in this direction is to obtain effective equations that incorporate singularity resolution and other discreteness corrections. The effect of these is already apparent in preliminary calculations[17] of temperature and specific heat corrections of black holes. These suggest that the end point of black hole evaporation is a Planck scale remnant. Similar arguments give a correction to the mass loss formula (9) that takes the form

$$\frac{dM}{dt} = -\alpha\,\frac{M_P}{t_P}\,M_P^2\left(\frac{\sqrt{|M+\lambda M_P|}-\sqrt{|M-\lambda M_P|}}{\lambda M_P}\right)^4, \tag{22}$$

where λ is a dimensionless discreteness constant. This formula comes from using the relationship between black hole mass and horizon radius, and the observation that fundamental discreteness results in a formula of the type (20) for an inverse radius operator.

4. Summary and outlook

We have seen the role played by fundamental discreteness in formulating a quantum theory of gravity, and its use in a model for gravitational collapse. The results so far from this approach to understanding black hole formation in quantum gravity are threefold: (i) a lattice based canonical quantization procedure, (ii) singularity free and unitary evolution equations, and (iii) a test for black holes in a quantum dynamical setting which makes no use of classical boundary conditions at event horizons.

The main challenge is to use the formalism to explicitly compute the evolution of a given matter-geometry state until it satisfies the quantum black hole criteria, and then to continue to the evolution to see if and how Hawking radiation might arise. An aspect of this work that incorporates quantum gravity corrections to the evolution equations has already given indications[18] that the approach to black hole formation is significantly different from that seen classically[4].

References

1. N. D. Birrell and P.C. W. Davies, *Quantum fields in curved spacetime* (Cambridge University Press, Cambridge (1984)).
2. R. M. Wald, *Quantum fields in curved spacetime and black hole physics* (University of Chicago Press, Chicago (1994)).
3. V. Mukhanov, *Physical foundations of cosmology* (CUP, Cambridge (2005)).
4. M. W. Choptuik, Phys. Rev. Lett. 70, 9 (1993).
5. C. Gundlach, Phys. Rept. **376** 339 (2003).
6. S. W. Hawking, Commun. Math. Phys. **43** (1975) 199.
7. V. Husain, Class. Quantum Grav. 5, 575 (1988).
8. V. Husain, J. Louko, O. Winkler, Phys. Rev. D76, 084002 (2007); gr-qc/07070273.
9. T. Thiemann, *Introduction to modern canonical quantum general relativity*, gr-qc/0110034; A. Ashtekar, J. Lewandowski, *Background independent quantum gravity: a status report*, gr-qc/0404018; L. Smolin, *The case for background independence* hep-th/0507235.
10. T. Thiemann, Class. Quant. Grav. **15**, 839(1998); gr-qc/9606089.
11. M. Bojowald, Phys. Rev. D **64** 084018 (2001); gr-qc/0105067.
12. V. Husain, O. Winkler, Phys. Rev. D **69**, 084016 (2004); gr-qc/0312094.
13. V. Husain, O. Winkler, Class. Quant. Grav. **22**, L127 (2005); gr-qc/0410125.
14. V. Husain, O. Winkler, Phys. Rev. D71 104001 (2005); gr-qc/0503031.
15. J. Thornburg, *Event and apparent horizon finders in 3+1 numerical relativity*, gr-qc/0512169.
16. V. Husain, O. Winkler, Class. Quantum Grav. **22**, L135 (2005).
17. V. Husain, O. Winkler, Int. J. Mod. Phys. D 14, 2233 (2005); gr-qc/0505153.
18. V. Husain, G. Kunstatter, to appear (2007).

Constraints on alternative theories of gravity and cosmology

Alexander F. Zakharov

National Astronomical Observatories of Chinese Academy of Sciences, 20A Datun Road, Chaoyang District, Beijing, 100012, China; Institute of Theoretical and Experimental Physics, B. Cheremushkinskaya, 25, 117259, Moscow, Russia; Center of Advanced Mathematics and Physics, National University of Sciences and Technology, Rawalpindi, Pakistan; Bogoliubov Laboratory for Theoretical Physics, JINR, 141980 Dubna, Russia

E-mail: zakharov@itep.ru

One could call 2006 as the year of cosmology since in that year two US scientists were awarded the Nobel prize for their studies of the Cosmic Microwave Background (CMB) spectrum and anisotropy. Here I recall earlier studies of the CMB anisotropy done with the Soviet spacecraft Prognoz-9 by the Relikt-1 team. Problems of modern cosmology are outlined. We discuss conformal cosmology parameters from supernovae data in brief. Two approaches to solve the basic problems of cosmology, such as dark matter and dark energy, are discussed, the first (standard) possibility is to introduce new particles, fields etc; the second possibility is to try to change the law of gravity to fit observational data. We discuss advantages and disadvantages of the second choice.

Keywords: Cosmology; Cosmological tests; Cosmic microwave background radiation; Space experiments

1. Introduction

For the scientific community 2005 was the World year of physics due to publications of the famous Einstein's papers and the birth of modern (contemporary) physics in 1905. The next year (2006) was also remarkable for the physics community (it was the year of astrophysics and cosmology) since in 2006 the Nobel prize for CMB studies was presented to J. Mather and G. Smoot. Moreover J. Mather and the COBE (COsmic Background Explorer) team were awarded the Peter Gruber prize on cosmology in the same year at the General Assembly of the International Astronomical Union in Prague. No doubt, 2007 will be recognized as the very important year for astrophysical (or probably for scientific in general) community since a number of of discoveries have been made even up to the present date. For Russian space science 2007 is a jubilee year since the first artificial satellite (Sputnik) was launched fifty years ago on October 4, 1957.

The structure of the paper is the following. In section 2 we recall results of CMB anisotropy studies done with the Soviet Relikt-1 mission. In section 3 we outline the current status of cosmological studies and parameters of conformal cosmological models in brief. In section 4 we discuss the success of an alternative way to solve the Dark Matter (DM) and Dark Energy (DE) problems (so-called geometrical approach or $f(R)$-models), where the law of gravity law is changed by changing the classical Lagrangian for gravity and severe constraints on parameters of the resulting so-called R^n theory from Solar system data are given. Conclusion are presented in section 5.

2. CMB anisotropy studies, Relikt-1 & COBE

Cosmic microwave background (CMB) existence was predicted in the framework of the the so called Big Bang cosmological model.[1] The CMB radiation was discovered by A. Penzias and R. Wilson[2] in an unexpected way (and they were awarded the Nobel prize for the discovery in 1978).

Because the Earth moves in respect to the CMB, a dipole temperature anisotropy is expected. A dipole temperature anisotropy of the level of $\Delta T/T = 10^{-3}$ was observed, which corresponds to a peculiar velocity 380 km/s of the Earth towards the constellation Virgo (first measurements of dipole anisotropy were not very precise,[3] but later the accuracy was significantly improved[4,5]).

2.1. *The Relikt-1 experiment*

In 1983, in the Soviet Union the Relikt-1 experiment[a] was conducted aboard the Prognoz-9 spacecraft[b] in order to investigate CMB radiation from space for the first time in history. As many other Prognoz missions, the scientific payload was prepared by the Space Research Institute of the Soviet Academy of Sciences. Dr. Igor Strukov was the principal investigator.

The spacecraft Prognoz-9, had an 8 mm band radiometer with an extremely high sensitivity of 35 μK per second and it was launched into a

[a]In Russian literature CMB radiation is known as relic radiation, therefore the name Relic ("Relikt" in Russian) was chosen.
[b]In Russian "Prognoz" means "forecast". Prognoz spacecrafts were designed originally for studies of Solar activity (Solar flares in particular) and its impact on geophysics, geomagnetic fields and space weather.[6] The spacecrafts were launched at high elliptical orbits with apogee about 200 000 km.

high apogee orbit with a 400,000 km semi-axis. The high orbit was a great advantage of the mission since it allowed one to reduce the impact of the geomagnetic field on measurements. A disadvantage of the experiment was that the observations were conducted only in one spectral band, therefore leaving a lot freedom for theoretical interpretations of the results, in contrast to which multi-band measurements provide very small room for alternative explanations of anisotropy.

The radiometer scanned the entire celestial sphere for six months. Computer facilities (and therefore, data analyzing) were relatively slow at the time. Preliminary analysis of anisotropy studies indicate upper limits on anisotropy,[10] but in this case even the negative results (upper limits) were extremely important to evaluate the sensitivity required to design detectors for the next missions (including COBE).

In 1986, at the Space Research Institute it was decided to prepare a next generation of space experiments to study the anisotropy of CMB, and to start to develop the Relikt-2 project. The sensitivity of the detectors was planned to be 20 times better than the Relikt-1 sensitivity. The Libris satellite was scheduled to carry the Relikt-2 payload and the spacecraft was planned to be located near the Lagrangian point L2 (in the Sun – Earth system).[c] Originally, there was a plan to launch the Libris spacecraft in 1993–1994. However, the project has not been realized, basically due to a lack of funds. To prepare Relikt-2, the team members re-analyzed Relikt-1 data and finally in beginning of 1991 they discovered signatures of the quadrupole anisotropy, but I. Strukov needed to check the conclusions again and again. The discovery of anisotropy by the Relikt-1 spacecraft was first reported officially in January 1992 at the Moscow astrophysical seminar and the Relikt-1 team submitted papers in Soviet Astronomy Letters and Monthly Notices of Royal Astronomical Society[8] (soon after the papers were published).

Relikt-1 results are described in an adequate way at the at the official NASA web-site http://lambda.gsfc.nasa.gov/product/relikt/[d] "... The Relikt Experiment Prognoz 9, launched on 1 July 1983 into a high-apogee (700,000 km) orbit, included the Relikt-1 experiment to investigate the

[c]The trajectories near the L_2 point are suitable for different space experiments; for example, the orbit was selected for the WMAP spacecraft (it is one of the most successful missions to study the CMB anisotropy and such an orbit is planned to be used for the future Millimetron cryogenic telescope (it is scheduled to be launched around 2016)).
[d]See, also, an article "Cosmic microwave background radiation" in Wikipedia, the free encyclopedia (http://en.wikipedia.org/wiki/Main_Page).

anisotropy of the CMB at 37 GHz, using a Dicke-type modulation radiometer. During 1983 and 1984 some 15 million individual measurements were made (with 10% near the galactic plane providing some 5000 measurements per point). The entire sky was observed in 6 months. The angular resolution was 5.5 degrees, with a temperature resolution of 0.6 mK. The galactic microwave flux was measured and the CMB dipole observed. A quadrupole moment was found between 17 and 95 microKelvin rms, with 90% confidence level. A map of most of the sky at 37 GHz is available..." (also, references,[8] given at the NASA web-site).

2.2. *The Nobel prize*

The Nobel Prize in Physics for 2006 was awarded to John Mather and George Smoot. The Royal Swedish Academy of Sciences issued the Press Release, dated 3 October 2006, on The Nobel Prize in Physics 2006.[e] The Royal Swedish Academy of Sciences has decided to award the Nobel Prize in Physics for 2006 jointly to John C. Mather (NASA Goddard Space Flight Center, Greenbelt, MD, USA), and George F. Smoot (University of California, Berkeley, CA, USA) "for their discovery of the blackbody form and anisotropy of the cosmic microwave background radiation". J. Mather was the project scientist of the COBE mission and was responsible for measurements of CMB spectrum (he was the Principle Investigator (PI) of the Far InfraRed Absolute Spectrometer (FIRAS) experiment) and measurements of CMB spectrum were rather successful. G. Smoot made a great contribution towards measurements of the dipole anisotropy[5] and the quadrupole one[9], since he was PI of the differential microwave radiometer (DMR) aboard the COBE satellite and found the quadrupole anisotropy with these facilities, but both anisotropies were discovered earlier by other people.

2.3. *Summary on the Quadrupole Anisotropy Discovery*

The discovery of anisotropy by the Relikt-1 spacecraft was first reported officially in January 1992 at the Moscow astrophysical seminar. Relikt-1 team submitted their paper in Soviet Astronomy Letters and Monthly Notices of Royal Astronomical Society on January 19, 1992 and February 3, 1992.

On April 21, 1992, G. Smoot and his co-authors, the head of DMR experiment aboard the COBE mission, submitted a paper in Astrophysical

[e] Information from the Nobel prize committee web-site http://www.kva.se.

Journal (Letters)[9] and on April 22, 1992, Smoot reported at a press conference about the discovery of the CMB anisotropy with the COBE satellite. After that the mass media reported these results as a major scienctific success. In 1992, COBE results about discovery of the CMB anisotropy were reported elsewhere. However there is no doubt that the COBE collaboration knew the results of the Relikt-1 team and even quoted their upper limit on the quadrupole anisotropy published in the paper.[10] Sometimes, results of both (COBE and Relikt-1) collaborations were presented at the same scientific meetings. In particular, in June 1992, at the conference on Particle Astrophysics in Blois (France) results on discoveries of CMB anisotropy were presented at a plenary session by representatives of both teams (Relikt-1 and COBE), namely M. V. Sazhin and G. Smoot. Smoot's paper was published in the conference proceedings,[11] but unfortunately, eventually, Sazhin's contribution was not published there.

However, summarizing, one could say that since papers of the Relikt-1 team were submitted on January 19, 1992 and February 3, 1992 in Soviet Astronomy and Monthly Notices of Royal Astronomy Society respectively, but COBE paper was submitted on April 21, 1992, one would conclude that the discovery of the quadrupole anisotropy was done by the Relikt-1 team and published in papers.[8]

3. Precise cosmology & Conformal Cosmological Models

Since 1998 people claim that cosmology has started to be a precise science. Really, if the standard cosmological paradigm was chosen, now we can evaluate cosmological parameters with a precision about 10 % (typically only statistical errors are indicated and it is assumed that systematical errors are negligible). However in the beginning of 1998 it was found that the Universe is open and $\Omega_m \sim 0.2 - 0.4$ (based on observations of galactic clusters,[12]) but at the end of 1998, it was concluded that the Universe is flat with $\Omega_m = 0.3, \Omega_\Lambda = 0.7$[13] (assuming that the SNe Ia are standard candles).

However, other alternatives are allowed apart from the standard cosmological approach; for example, a conformal cosmological (CC) model was discussed.[14] Using "gold" and "silver" 186 SNe Ia[15] we confirm in general and clarify previous conclusions about CC model parameters,[16] done earlier with analysis of smaller sample of SNe Ia data[14] that the pure flat rigid CC model could fit the data relatively well since $\Delta\chi^2 \approx 44.3$ (or less than 20 %) in respect of the standard cosmology flat model with $\Omega_m = 0.28$. Other pure flat CC models should be ruled out since their χ^2 values are too high.

Thus, for CC model fits calculated with SNe Ia data, in some sense, a rigid equation of state could substitute the Λ-term (or quintessence) in the Universe. As was mentioned in the paper[16] rigid matter can be formed by a free massless scalar field. The best CC models provide almost the same quality fits of SNe Ia data as the best fit for the Standard Cosmological (CS) flat model. However the last (generally accepted) model is more preferable.

4. Standard Cosmology vs. f(R) gravity

The standard cosmological model has problems in understanding the origin of Dark Matter (DM) and dark energy (DE). Another approach was proposed and it was based on an assumption that gravity is different from standard general relativity and that it can be described by a modified Lagrangian.[17] The model was successful in explaining the acceleration of the Universe, but has problems to fit Solar system data. Recently, a generalization of the approach was proposed in the framework of higher order theories of gravity – also referred to as $f(R)$ (or fourth order) theories. This approach proposes a modification of the gravity action with the form[18]

$$\mathcal{A} = \int d^4x\sqrt{-g}[f(R) + \mathcal{L}_m], \tag{1}$$

where $f(R)$ is a generic function of the Ricci scalar curvature and $\mathcal{L}_m$ is the standard matter Lagrangian. For example, if $f(R) = R + 2\Lambda$ the theory coincides with General Relativity (GR) with the Λ term. In particular, power law function $f(R)$ theories of the form $f(R) = f_0 R^n$ were considered.

These kinds of theories were very successful in explaining standard cosmological data such as SNe Ia fits, the acceleration of the Universe[18], rotation curves for galaxies[19]. It was suggested that standard general relativity plus DM and DE may be distinguished from R^n approaches with gravitational microlensing[20], but Solar system data (planetary orbital periods, in particular) already put severe constraints on parameters of the the theories.[21]

5. Conclusions

In conclusion we note that, unfortunately, even well-informed Russian (and other) authors do not cite Relikt-1 results in papers on cosmology, where the COBE anisotropy result is quoted as the only experiment that discovered the phenomenon. It means that the Nobel prize winner (1978) P. L. Kapitza[f] was correct when he wrote about this kind of problem in 1946: "... Our main national defect is an underestimation of our powers and overestimation of foreign ones. So, an extra modesty is much more defective than an extra self-confidence... Very often a cause of unused innovations is that usually we underestimate our own discoveries and overestimate foreign ones..."

For CC model fits calculated with SNe Ia data, in some sense, a rigid equation of state could substitute the Λ-term (or quintessence) in the Universe. Rigid matter can be formed by a free massless scalar field.[16]

Solar system constraints put severe constraints on parameters of alternative theories of gravity.[21]

Acknowledgements

The author thanks Prof. Riazuddin for his kind attention to this contribution, and F. De Paolis, G. Ingrosso, A. A. Nucita, V. N. Pervushin, A. Qadir for fruitful discussions. AFZ is grateful to the National Natural Science Foundation of China (NNSFC) (Grant # 10233050) and National Basic Research Program of China (2006CB806300) for a partial financial support of the work.

References

1. G. Gamov, *Phys. Rev.* **74**, 505 (1948); G. Gamov, *Nature* **162**, 680 (1948); R. A. Alpher, R. Herman, *Phys. Rev.* **74**, 1577 (1948).
2. A. A. Penzias, R. W. Wilson, *Astrophys. J.* **142**, 419 (1965).
3. E. K. Conklin, *Nature*, **222**, 971 (1969); P. S. Henry, *Nature*, **231**, 516 (1971).
4. B. E. Corey, D. T. Wilkinson, *Bull. Amer. Astron. Soc.*, **8**, 351 (1976).
5. G.F. Smoot, M. V. Gorenstein, R. A. Muller, *Phys. Rev. Lett.*, **39**, 898 (1977).
6. Problems of Solar activity and space system "Prognoz", eds. S. N. Vernov et al., (Nauka, Moscow, 1977, in Russian).
7. A. A. Klypin, M. V. Sazhin, I. A. Strukov, D. P. Skulachev, M. V. Sazhin, *Sov. Astron. Lett.* **13**, 104 (1987).
8. I. A. Strukov, A. A. Brukhanov, D. P. Skulachev, M. V. Sazhin, *Sov. Astron. Lett.* **18**, 153 (1992); I. A. Strukov, A. A. Brukhanov, D. P. Skulachev, M. V. Sazhin, *Mon. Not. R. Astron. Soc.*, **258**, 37 (1992).

[f]He was awarded the Nobel prize for his experimental discovery of superfluidity and received the prize in the same year as Penzias and Wilson. He was Rutherford's favourite student and worked for a long time in the UK and therefore knew Western and Russian mentalities very well.

9. G.F. Smoot, C. L. Bennett, A. Kogut et al., *Astrophys. J.*, **396**, L1 (1992).
10. A.A. Klypin, M. V. Sazhin, I. A. Strukov, D. P. Skulachev, *Sov. Astron. Lett.* **13**, 104 (1992).
11. G.F. Smoot, in *Particle Astrophysics, Proc. of the 4th Rencontres de Blois*, Eds. G. Fontaine, J. Tran Thanh Van (Gif-sur-Yvette, Editions Frontieres, 1993) p. 409.
12. N.A. Bahcall, X. Fan, *Publ. Nat. Academy of Science*, **95**, 5956 (1998); N.A. Bahcall, X. Fan, *Astrophys. J.* **504**, 1 (1998).
13. A.G. Riess et al., *Astron. J.* **116**, 1009 (1998); S. Perlmutter et al., *Astrophys. J.* **517**, 565 (1999).
14. D. Behnke, D. B. Blaschke, V. N. Pervushin, D. V. Proskurin, *Phys. Lett. B* **530**, (2002). D. B. Blaschke, S. I. Vinitsky, A. A. Gusev, V. N. Pervushin, D. V. Proskurin, *Phys. Atom. Nucl.* **67**, 1050 (2004); B. M. Barbashov, V. N. Pervushin, A. F. Zakharov, V. A. Zinchuk, *Int. J. Mod. Phys. A* **21**, 5957 (2006); B. M. Barbashov, V. N. Pervushin, A. F. Zakharov, V. A. Zinchuk, *Physics Letters B* **633**, 438 (2006).
15. A.D. Riess, L.-G. Strolger, J. Tonry et al., *Astrophys. J.*, **607**, 665 (2004).
16. A. F. Zakharov, A. A. Zakharova, V. N. Pervushin, astro-ph/0611657
17. S. M. Carroll, V. Duvvuri, M. Trodden, M. S. Turner, *Phys. Rev. D* **70**, 043528 (2004).
18. S. Capozziello, R. Garattini, *Class. Quant. Grav.* **24**, 1627 (2007); S. Capozziello, V. F. Cardone, A. Troisi, *Phys. Rev. D* **73**, 104019 (2006); S. Capozziello, S. Nojiri, S. D. Odintsov, A. Troisi, *Phys. Lett. B* **639**, 135 (2006); S. Capozziello, V. F. Cardone, A. Troisi, *Mon. Not. Roy. Astron. Soc.* **375**, 1423 (2007); S. Capozziello, V. F. Cardone, A. Troisi, *J. Cosm. Astropart.* **8**, 1 (2006); S. Capozziello, S. Nojiri, S. D. Odintsov, *Phys. Lett. B* **634**, 93 (2006); S. Capozziello, V. F. Cardone, E. Elizalde, S. Nojiri, S. D. Odintsov, *Phys. Rev. D* **73**, 043512 (2006); S. Capozziello, A. Troisi, *Phys. Rev. D* **72**, 044022 (2005); S. Capozziello, V. F. Cardone, E. Piedipalumbo, C. Rubano, *Class. Quant. Grav.* **23**, 1205 (2006); S. Capozziello, V. F. Cardone, A. Troisi, *Phys. Rev. D* **71**, 043503 (2005); S. Capozziello, V. F. Cardone, M. Francaviglia, *Gen. Rel. Grav.* **38**, 711 (2006); S. Carloni, P. K. S. Dunsby, S. Capozziello, A. Troisi, *Class. Quant. Grav.* **22**, 4839 (2005); A. Borowiec, W. Godlowski, M. Szydlowski, *Int. J. Geom. Meth. Mod. Phys.* **4**, 183 (2007); A. Borowiec, W. Godlowski, M. Szydlowski, *Phys. Rev. D* **74**, 043502 (2006).
19. S. Capozziello, V. F. Cardone, A. Troisi, *Mon. Not. Roy. Astron. Soc.* **375** 1423 (2007); C. F. Martins, P. Salucci, astro-ph/0703243.
20. S. Capozziello, V. F. Cardone, A. Troisi, *Phys. Rev. D* **73**, 104019 (2006).
21. A. F. Zakharov, A. A. Nucita, F. De Paolis, G. Ingrosso, *Phys. Rev. D* **74**, 107101 (2006).

International Symposium on Contemporary Physics
March 26 – 30 2007

List of Participants

Abbas, Tahir
Bahauddin Zakariya University
Multan, Pakistan

Abbas, Tasawar
Quaid-i-Azam University
Islamabad, Pakistan

Abdullah, Ali
Government College University
Lahore, Pakistan

Afgan, Sabeen Sher
Quaid-i-Azam University
Islamabad, Pakistan

Afzal, Muhammad
Optics Laboratories Nilore
Islamabad, Pakistan

Ahkam, Qazi Muhammad
COMSATS Institute of Information Technology
Islamabad, Pakistan

Ahmad, Ahsan
Quaid-i-Azam University
Islamabad, Pakistan

Ahmad, Ali
Pakistan Institute of Nuclear Science and Technology
Islamabad, Pakistan

Ahmad, Farooq
Punjab University
Lahore, Pakistan

Ahmad, Ishaq
COMSATS Institute of Information Technology
Islamabad, Pakistan

Ahmad, Mashkoor
Pakistan Institute of Nuclear Science and Technology
Islamabad, Pakistan

Ahmad, Masud
Pakistan Atomic Energy Commission
Islamabad, Pakistan

Ahmad, Muhammad
Quaid-i-Azam University
Islamabad, Pakistan

Ahmad, Mushtaq
Pakistan Institute of Nuclear Science and Technology
Islamabad, Pakistan

Ahmed, Ijaz
National Centre for Physics
Islamabad, Pakistan

Ahmed, Ishtiaq
National Centre for Physics
Islamabad, Pakistan

Ahmed, Kamaluddin
COMSATS Institute of Information Technology
Islamabad, Pakistan

Ahmed, Rizwan
Quaid-i-Azam University
Islamabad, Pakistan

Ahmed, Sarfraz
Pakistan Atomic Energy Commission
Islamabad, Pakistan

Ahmed, Zahoor
Quaid-i-Azam University
Islamabad, Pakistan

Ajaib, Muhammad Adeel
Quaid-i-Azam University
Islamabad, Pakistan

Akbar, Sadaf
Quaid-i-Azam University
Islamabad, Pakistan

International Symposium on Contemporary Physics
March 26 – 30 2007

List of Participants

Akhtar, Muhammad Jawad
Pakistan Institute of Nuclear Science and Technology
Islamabad, Pakistan

Ali, Abid
Optics Laboratories Nilore
Islamabad, Pakistan

Ali, Ahmed
Deutsches Elektronen Synchrotron
Hamburg, Germany

Ali, Akbar
Quaid-i-Azam University
Islamabad, Pakistan

Ali, Junaid
COMSATS Institute of Information Technology
Islamabad, Pakistan

Ali, Muhammad Mazhar
Quaid-i-Azam University
Islamabad, Pakistan

Ali, Muneeb
Khan Research Laboratories
Islamabad, Pakistan

Ali, Raheel
Quaid-i-Azam University
Islamabad, Pakistan

Ali, Sajid
National University of Science and Technology
Islamabad, Pakistan

Ali, Shoukat
Pakistan Institute of Nuclear Science and Technology
Islamabad, Pakistan

Ali, Zulifqar
Optics Laboratories Nilore
Islamabad, Pakistan

Altaf ul Haq, Muhammad
University of the Punjab
Lahore, Pakistan

Amjad, Asma
Quaid-i-Azam University
Islamabad, Pakistan

Ammar, Shahzada Tasir
Quaid-i-Azam University
Islamabad, Pakistan

Anis-Ur-Rehman, Muhammad
COMSATS Institute of Information Technology
Islamabad, Pakistan

Anjum, Naveed
Optics Laboratories Nilore
Islamabad, Pakistan

Anjum, Safia
University of Engineering & Technology
Lahore, Pakistan

Ansari, Muhammad Hamid
National Centre for Physics
Islamabad, Pakistan

Anwar, Hafeez
Quaid-i-Azam University
Islamabad, Pakistan

Anwar, Javaid
COMSATS Institute of Information Technology
Islamabad, Pakistan

Anwar, Mamuna
Quaid-i-Azam University
Islamabad, Pakistan

Arif, Muhammad
Quaid-i-Azam University
Islamabad, Pakistan

International Symposium on Contemporary Physics
March 26 – 30 2007

List of Participants

Arkwincheev, Valriy E.
Federal Urdu University
Karachi, Pakistan

Arshad, Muhammad
National Centre for Physics
Islamabad, Pakistan

Arshad, Nigum
Quaid-i-Azam University
Islamabad, Pakistan

Ashraf, Imrana
Quaid-i-Azam University
Islamabad, Pakistan

Ashraf, Mian Muhammad
Optics Laboratories Nilore
Islamabad, Pakistan

Asif, Muhammad
Pakistan Institute of Nuclear Science and Technology
Islamabad, Pakistan

Aslam, Jamil
National Centre for Physics
Islamabad, Pakistan

Aslam, Muhammad
Pakistan Atomic Energy Commission
Islamabad, Pakistan

Aslam, Muhammad Naveed
Government College University
Lahore, Pakistan

Ata-Ul-Hakim
Quaid-i-Azam University
Islamabad, Pakistan

Aurangzeb
Quaid-i-Azam University
Islamabad, Pakistan

Awan, Saifullah
Government College University
Lahore, Pakistan

Ayub, Muhammad
Government College University
Lahore, Pakistan

Azim, Tasnim
Quaid-i-Azam University
Islamabad, Pakistan

Aziz, Uzma
Pakistan Institute of Nuclear Science and Technology
Islamabad, Pakistan

Baig, Aslam
Quaid-i-Azam University
Islamabad, Pakistan

Batool, Nazia
Quaid-i-Azam University
Islamabad, Pakistan

Bergou, Janos
City University of New York
New York, U.S.A.

Bhatti, Arshad Saleem
COMSATS Institute of Information Technology
Islamabad, Pakistan

Bhatti, Khurshid Aslam
University of Engineering & Technology
Lahore, Pakistan

Butt, Aatif Imtiaz
Pakistan Atomic Energy Commission
Islamabad, Pakistan

Butt, Jamila
National Centre for Physics
Islamabad, Pakistan

International Symposium on Contemporary Physics
March 26 – 30 2007

List of Participants

Chohan, Muhammad Islam
Khan Research Laboratories
Islamabad, Pakistan

Ehsan, Saqib
National Centre for Physics
Islamabad, Pakistan

Elahi, Irfan
University of Agriculture
Faisalabad, Pakistan

Faiz, Sumaira
Bahauddin Zakariya University
Multan, Pakistan

Farooq, Muhammad Asif
Quaid-i-Azam University
Islamabad, Pakistan

Fatima, Sadaf
University of Karachi
Karachi, Pakistan

Fayyazuddin
National Centre for Physics
Islamabad, Pakistan

Firdous, Tayyaba
Quaid-i-Azam University
Islamabad, Pakistan

Firdous, Umber
Quaid-i-Azam University
Islamabad, Pakistan

Fromherz, Peter
Max Planck Institute for Biochemistry
Martinsried, Germany

Ghafoor, Fazal
COMSATS Institute of Information Technology
Islamabad, Pakistan

Gilani, Amjad Hussain Shah
National Centre for Physics
Islamabad, Pakistan

Gul, Anas
Pakistan Atomic Energy Commission
Islamabad, Pakistan

Habib, Irfan
National University of Science and Technology
Islamabad, Pakistan

Hafeez, Sarwat
Quaid-i-Azam University
Islamabad, Pakistan

Hamdani, Ajmal H.
Optics Laboratories Nilore
Islamabad, Pakistan

Haseeb, Mahnaz
COMSATS Institute of Information Technology
Islamabad, Pakistan

Hasnain, Haseeb
Pakistan Institute of Nuclear Science and Technology
Islamabad, Pakistan

Hassan, Mehwish
Quaid-i-Azam University
Islamabad, Pakistan

Hasanain, S. K
Quaid-i-Azam University
Islamabad, Pakistan

Hoorani, Hafeez R.
National Centre for Physics
Islamabad, Pakistan

Hormes, Franz-Josef
Louisiana State University,
Baton Rouge, Louisiana , U.S.A.

International Symposium on Contemporary Physics
March 26 – 30 2007

List of Participants

Husain, Altaf
University of Karachi
Karachi, Pakistan

Hussain, Ashfaq Khosa
COMSATS Institute of Information Technology
Islamabad, Pakistan

Hussain, Ghazanfar
Pakistan Atomic Energy Commission
Islamabad, Pakistan

Hussain, Ibrar
National University of Science and Technology
Islamabad, Pakistan

Hussain, Irshad
National Institute of Biotechnology and Genetic Engineering
Faisalabad, Pakistan

Hussain, Muhammad Asghar
Islamia University
Bahawalpur, Pakistan

Hussain, Naveed
COMSATS Institute of Information Technology
Islamabad, Pakistan

Hussain, Rafaqat
COMSATS Institute of Information Technology
Islamabad, Pakistan

Hussain, Safdar
University of Sargodha
Sargodha, Pakistan

Hussain, Shahid
Quaid-i-Azam University
Islamabad, Pakistan

Hussain, Shahzad
Quaid-i-Azam University
Islamabad, Pakistan

Hussain, Syed Asad
University of Karachi
Karachi, Pakistan

Hussain, Talib
Khan Research Laboratories
Islamabad, Pakistan

Husain, Viqar
University of New Brunswick
Fredericton, New Brunswick, Canada.

Ikram, Manzoor
COMSATS Institute of Information Technology
Islamabad, Pakistan

Imran, Zafar
Quaid-i-Azam University
Islamabad, Pakistan

Inam-Ur-Rehman
Quaid-i-Azam University
Islamabad, Pakistan

Iqbal, Muhammad
Pakistan Institute of Nuclear Science and Technology
Islamabad, Pakistan

Iqbal, Muhammad Zafar
Quaid-i-Azam University
Islamabad, Pakistan

Iqbal, Shahid
Quaid-i-Azam University
Islamabad, Pakistan

Iqbal, Tariq
Quaid-i-Azam University
Islamabad, Pakistan

International Symposium on Contemporary Physics
March 26 – 30 2007

List of Participants

Islam, Muhammad
Govt. Degree College
Rawalpindi, Pakistan

Israr, Zeba
Government College University
Lahore, Pakistan

Jabbar, Abdul
Pakistan Institute of Nuclear Science and Technology
Islamabad, Pakistan

Jabeen, Maria
Quaid-i-Azam University
Islamabad, Pakistan

Jahangir, Rabia
Quaid-i-Azam University
Islamabad, Pakistan

Jameel, Abdul
COMSATS Institute of Information Technology
Islamabad, Pakistan

Jamil, Mubasher
National University of Science and Technology
Islamabad, Pakistan

Janjua, Sohail Ahmad
Pakistan Institute of Nuclear Science and Technology
Islamabad, Pakistan

Jehan, Nusrat
Quaid-i-Azam University
Islamabad, Pakistan

Johnson, Louise
University of Oxford
Oxford, UK

Junaid, Muhammad
COMSATS Institute of Information Technology
Islamabad, Pakistan

Jürgen, Körner.
Johannes Gutenberg University of Mainz
Mainz, Germany

Khan, Ehsanullah.
COMSATS Institute of Information Technology
Islamabad, Pakistan

Khan, Faizullah
Gomal, University
D. I. Khan, Pakistan

Khan, Fazeel Mahmood
Government College University
Lahore, Pakistan

Khan, Gulfam
Pakistan Atomic Energy Commission
Islamabad, Pakistan

Khan, Jehan Zeb Alam
Quaid-i-Azam University
Islamabad, Pakistan

Khan, Liaqat Ali
Quaid-i-Azam University
Islamabad, Pakistan

Khan, M. Yaqub
Quaid-i-Azam University
Islamabad, Pakistan

Khan, Mahtab Ahmad
Quaid-i-Azam University
Islamabad, Pakistan

Khan, Muhammad Afzal
Air Weapon Complex
Wah Cantt. Pakistan

Khan, Muhammad Ayub
University of Karachi
Karachi, Pakistan

International Symposium on Contemporary Physics
March 26 – 30 2007

List of Participants

Khan, Muhammad Khalid
Quaid-i-Azam University
Islamabad, Pakistan

Khan, Muhammad Nasir
Pakistan Institute of Nuclear Science and Technology
Islamabad, Pakistan

Khan, Nawazish Ali
Quaid-i-Azam University
Islamabad, Pakistan

Khan, Rashid
Gomal University
D. I. Khan, Pakistan

Khan, Riaz
Pakistan Institute of Nuclear Science and Technology
Islamabad, Pakistan

Khan, Rizwan Ahmad
COMSATS Institute of Information Technology
Islamabad, Pakistan

Khan, Shabbir Ahmad
COMSATS Institute of Information Technology
Islamabad, Pakistan

Khan, Umar Saeed
University of Peshawar
Peshawar, Pakistan

Khan, Wajid Ali
Quaid-i-Azam University
Islamabad, Pakistan

Khattak, Nazir Shah
University of Peshawar
Peshawar, Pakistan

Khattak, Noor Abbas Din
Gomal University
D. I. Khan, Pakistan

Khokhar, Sarfraz Ahmed
Pakistan Atomic Energy Commission
Islamabad, Pakistan

Kiran, Zobia
Government College University
Lahore, Pakistan

Klitzing, K. Von
Max Planck Institute for Solid State Research
Stuttgart, Germany

Lutschak, William
New Mexico Academy
Mexico

Maartens, Roy
University of Portsmouth
Portsmouth, UK

Maaz
Pakistan Institute of Nuclear Science and Technology
Islamabad, Pakistan

Mahmood, Arshad
Pakistan Institute of Nuclear Science and Technology
Islamabad, Pakistan

Mahmood, Muhammad
Quaid-i-Azam University
Islamabad, Pakistan

Mahmood, Nasir
Optics Laboratories Nilore
Islamabad, Pakistan

Mahmood, Shahzad
Pakistan Institute of Nuclear Science and Technology
Islamabad, Pakistan

Mahmood, Shoukat
University of Azad Jammu and Kashmir
Muzaffarabad, Pakistan

International Symposium on Contemporary Physics
March 26 – 30 2007

List of Participants

Majid, Abdul
University of Azad Jammu and Kashmir
Muzaffarabad, Islamabad

Majid, Abdul
COMSATS Institute of Information Technology
Islamabad, Pakistan

Maken, Mukhtar H.
Khan Research Laboratories
Islamabad, Pakistan

Malik, Asim Ali
Government College University
Lahore, Pakistan

Malik, Muhammad Imran
Quaid-i-Azam University
Islamabad, Pakistan

Mann, Amer Bilal
Quaid-i-Azam University
Islamabad, Pakistan

Manzoor, Sadia
COMSATS Institute of Information Technology
Islamabad, Pakistan

Manzoor, Syed Asad
Quaid-i-Azam University
Islamabad, Pakistan

Manzoor, Umair
COMSATS Institute of Information Technology
Islamabad, Pakistan

Maqsood, Asghari
Quaid-i-Azam University
Islamabad, Pakistan

Maqsood, Muhammad
Khan Research Laboratories
Islamabad, Pakistan

Masood, Waqas
Pakistan Institute of Nuclear Science and Technology
Islamabad, Pakistan

Mehfooz, Shakeel
Quaid-i-Azam University
Islamabad, Pakistan

Mehmood, Shahid
Quaid-i-Azam University
Islamabad, Pakistan

Mehmood, Tariq
Khan Research Laboratories
Islamabad, Pakistan

Mehmood, Waqas
COMSATS Institute of Information Technology
Islamabad, Pakistan

Mir, Azeem
COMSATS Institute of Information Technology
Islamabad, Pakistan

Mirza, Arshad Majeed
Quaid-i-Azam University
Islamabad, Pakistan

Mirza, Imran Majid
Quaid-i-Azam University
Islamabad, Pakistan

Mirza, Mueed Ahmad
Government College University
Lahore, Pakistan

Mohammad, Nek
Quaid-i-Azam University
Islamabad, Pakistan

Monir, Faiqah Bint
Air Weapon Complex
Wah Cantt, Pakistan

International Symposium on Contemporary Physics
March 26 – 30 2007

List of Participants

Mufti, Tajdar
University of Karachi
Karachi, Pakistan

Mukhtar, Qaisar
Quaid-i-Azam University
Islamabad, Pakistan

Mumtaz, Arif
Quaid-i-Azam University
Islamabad, Pakistan

Muneeb-ur-Rahman
Ghulam Ishaq Khan Institute of Engineering Sciences and Technology
Swabi, Pakistan

Nadeem, Muhammad
Quaid-i-Azam University
Islamabad, Pakistan

Nadeem, Muhammad
Pakistan Institute of Nuclear Science and Technology
Islamabad, Pakistan

Nadeem, Muhammad Younus
Bahauddin Zakariya University
Multan, Pakistan

Nadeem, Mustansar
Quaid-i-Azam University
Islamabad, Pakistan

Naeem, Ismat
Quaid-i-Azam University
Islamabad, Pakistan

Naeem, Muddassar
Quaid-i-Azam University
Islamabad, Pakistan

Najam, Rabia
University of Karachi
Karachi, Pakistan

Naseer, Khalid
Quaid-i-Azam University
Islamabad, Pakistan

Naseer, Muhammad Kashif
COMSATS Institute of Information Technology
Islamabad, Pakistan

Naveed, Ahmad
Quaid-i-Azam University
Islamabad, Pakistan

Naveed, Muhammad
Optics Laboratories Nilore
Islamabad, Pakistan

Nawaz, Ahmad
Quaid-i-Azam University
Islamabad, Pakistan

Nawaz, Shahid
Pakistan Atomic Energy Commission
Islamabad, Pakistan

Nazir, Tahir
COMSATS Institute of Information Technology
Islamabad, Pakistan

Newman, Harvey.
California Institute of Technology
Pasadena, California, U.S.A.

Niraula, Om Prakash
Tribhuvan University
Kritipur, Kathmandu, Nepal

Nisar, Nadeem
Optics Laboratories Nilore
Islamabad, Pakistan

International Symposium on Contemporary Physics
March 26 – 30 2007

List of Participants

Noman, Muhammad
Pakistan Atomic Energy Commission
Islamabad, Pakistan

Nasim-ul-Haq, Mirza
Pakistan Institute of Nuclear Science and Technology
Islamabad, Pakistan

Nuzhat
Quaid-i-Azam University
Islamabad, Pakistan

Paracha, Muhammad Ali
National Centre for Physics
Islamabad, Pakistan

Petrova, Anna
National University of Science and Technology
Islamabad, Pakistan

Qadir, Asghar
National University of Science and Technology
Islamabad, Pakistan

Qamar-Ul-Haque
Pakistan Institute of Nuclear Science and Technology
Islamabad, Pakistan

Qamar-ul-Wahab
COMSATS Institute of Information Technology
Islamabad, Pakistan

Qasim, Irfan
Quaid-i-Azam University
Islamabad, Pakistan

Qasir, Maqsood-ul-Hasan
Pakistan Atomic Energy Commission
Islamabad, Pakistan

Qazi, Jawad
Harvard University
Boston, Massachusetts, U.S.A.

Qureshi, Nazar Abbas
Quaid-i-Azam University
Islamabad, Pakistan

Qureshi, Umer Saeed
Quaid-i-Azam University
Islamabad, Pakistan

Qurrat-Ul-Ain
COMSATS Institute of Information Technology
Islamabad, Pakistan

Rafiq, Muhammad
Quaid-i-Azam University
Islamabad, Pakistan

Rameez-ul-Islam
Quaid-i-Azam University
Islamabad, Pakistan

Ramzan, Muhammad
Quaid-i-Azam University
Islamabad, Pakistan

Rao, Saleem
National Centre for Physics
Islamabad, Pakistan

Rashid, Muneer Ahmed
National University of Science and Technology
Islamabad, Pakistan

Rawat, Abhishek
Inter University Centre for Astronomy and Astrophysics,
Pune, India

Raza, M. Waseem Khan
Centre for Earthquake Studies
Islamabad, Pakistan

International Symposium on Contemporary Physics
March 26 – 30 2007

List of Participants

Raza, Qaiser
Pakistan Institute of Nuclear Science and Technology
Islamabad, Pakistan

Raza, Syed Shabbar
National Centre for Physics
Islamabad, Pakistan

Rehman, Shama
Quaid-i-Azam University
Islamabad, Pakistan

Rehman, Talat
University of Central Florida,
Orlando, Florida, U.S.A.

Riazuddin
National Centre for Physics
Islamabad, Pakistan

Richardson, Martin
University of Central Florida,
Orlando, Florida, U.S.A.

Rind, Muhammad Akhtar
University of Sindh
Jamshoro, Pakistan

Rizvi, Sajjad
National University of Science and Technology
Islamabad, Pakistan

Rizvi, Tasneem Zahra
Quaid-i-Azam University
Islamabad, Pakistan

Sabeeh, Kashif
Quaid-i-Azam University
Islamabad, Pakistan

Saeed, Usman
Quaid-i-Azam University
Islamabad, Pakistan

Saeed-Ur-Rehman
COMSATS Institute of Information Technology
Islamabad, Pakistan

Saeed-Ur-Rehman
Pakistan Institute of Nuclear Science and Technology
Islamabad, Pakistan

Saif, Farhan
Quaid-i-Azam University
Islamabad, Pakistan

Saif-ur-Rehman
National Institute of Biotechnology and Genetic
Engineering
Faisalabad, Pakistan

Sajid, Muhammad
Pakistan Institute of Nuclear Science and Technology
Islamabad, Pakistan

Sajjad, Muhammad
Ghulam Ishaq Khan Institute of Engineering
Sciences and Technology,
Swabi, Pakistan

Saleem, Hamid
Pakistan Institute of Nuclear Science and Technology
Islamabad, Pakistan

Saleem, Muhammad
Optics Laboratories Nilore
Islamabad, Pakistan

Sarwar, Muhammad Adnan
Quaid-i-Azam University
Islamabad, Pakistan

Sattar, Abdul
University of Engineering and Technology
Lahore, Pakistan

International Symposium on Contemporary Physics
March 26 – 30 2007

List of Participants

Schleich, P. Wolfgang
University of Ulm
Ulm, Germany

Schopper, Herwig.
European Organization for Nuclear Research,
Geneva, Switzerland

Shabbir, Babar
Quaid-i-Azam University
Islamabad, Pakistan

Shabbir, Ghulam
Pakistan Institute of Nuclear Science and Technology
Islamabad, Pakistan

Shad, Mazher
Government College University
Lahore, Pakistan

Shafi, Aamir
National University of Science and Technology
Islamabad, Pakistan

Shafi-Ur-Rehman
Air Weapon Complex
Wah Cantt, Pakistan

Shah, Ismat
University of Delaware
Newark, Delaware, U.S.A.

Shah, K. Tahir
The Abdus Salam International Centre for Theoretical Physics,
Trieste, Italy.

Shaheen, Rubina
Pakistan Institute of Nuclear Science and Technology
Islamabad, Pakistan

Shahzad, M. Ikram
Pakistan Institute of Nuclear Science and Technology
Islamabad, Pakistan

Shakoor, Abdul
Quaid-i-Azam University
Islamabad, Pakistan

Sheikh, Muhammad Saleem
COMSATS Institute of Information Technology
Islamabad, Pakistan

Sher, Gul
Pakistan Institute of Nuclear Science and Technology
Islamabad, Pakistan

Shukla, Padma Kant
Ruhr-Universität Bochum
Bochum, Germany

Siddiqi, Irfan
University of California
Berkeley, UK

Siddique, Muhammad
Pakistan Institute of Nuclear Science and Technology
Islamabad, Pakistan

Smirnov, A.
The Abdul Salam International Centre for Theoretical Physics,
Trieste, Italy.

Sohaib, Muhammad Umer
COMSATS Institute of Information Technology
Islamabad, Pakistan

Sohail, Muhammad
Pakistan Atomic Energy Commission
Islamabad, Pakistan

International Symposium on Contemporary Physics
March 26 – 30 2007

List of Participants

Soomro, Kamran
National University of Science and Technology
Islamabad, Pakistan

Sreenivasan, K. R.
The Abdus Salam International Centre for Theoretical Physics,
Trieste, Italy

Suleymanov, Mais
COMSATS Institute of Information Technology
Islamabad, Pakistan

Sultan, Muhammad
National Centre for Physics
Islamabad, Pakistan

Tabassam, Hajrah
National Centre for Physics
Islamabad, Pakistan

Tagaymurod, Salikhov
Kohat University of Science and Technology
Kohat, Pakistan

Tahir, Nadeem
National Centre for Physics
Islamabad, Pakistan

Taj, Safia
National University of Science and Technology
Islamabad, Pakistan

Tiwana, Mazhar Hussain
Quaid-i-Azam University
Islamabad, Pakistan

Tonelli, Guido
European Organization for Nuclear Research,
Geneva, Switzerland

Toor, Abdul Hameed
Quaid-i-Azam University
Islamabad, Pakistan

Umair, Muhammad
Quaid-i-Azam University
Islamabad, Pakistan

Usman, Muhammad
Quaid-i-Azam University
Islamabad, Pakistan

Usman, Zahid
Gomal University
D. I. Khan, Pakistan

Waraich, Muhammad Ahmad
Punjab University
Lahore, Pakistan

Waris, Muhammad
Optics Laboratories Nilore
Islamabad, Pakistan

Wazir, Zafar
COMSATS Institute of Information Technology
Islamabad, Pakistan

Winick, Herman
Stanford University SLAC / SSRL
Stanford, USA

Yaqoob, Muhammad
Quaid-i-Azam University
Islamabad, Pakistan

Yaqub, Muhammad
Air Weapon Complex
Wah Cantt, Pakistan

Yasir, Muhammad
Quaid-i-Azam University
Islamabad, Pakistan

International Symposium on Contemporary Physics
March 26 – 30 2007

List of Participants

Yoshida, Zensho
University of Tokyo,
Tokyo, Japan

Younas, Waqar
Quaid-i-Azam University
Islamabad, Pakistan

Younis, Adnan
Quaid-i-Azam University
Islamabad, Pakistan

Zaidi, Syed Hassan Ali Shah
Gomal University
D. I. Khan, Pakistan

Zakaullah, Muhammad
Quaid-i-Azam University
Islamabad, Pakistan

Zakharov, Alexander
National University of Science and Technology
Islamabad, Pakistan

Zaman, Haider
Quaid-i-Azam University
Islamabad, Pakistan

Zaman, Sameena Shah
Quaid-i-Azam University
Islamabad, Pakistan

Zeb, Muhammad Ahsan
Quaid-i-Azam University
Islamabad, Pakistan

Zia, Muhammad Aslam
Optics Laboratories Nilore
Islamabad, Pakistan

Zubairy, M. Suhail
Texas A& M University
College Station, Texas, U.S.A